W9-AFB-602

523.109
F41r

130560

DATE DUE			
SEP 2 5 1985			

DISCARDED

CARL A. RUDISILL LIBRARY
LENOIR-RHYNE COLLEGE
HICKORY, NC 28603

THE RED LIMIT

The Search for the Edge of the Universe

TIMOTHY FERRIS

Second Edition: REVISED AND UPDATED

Introduction by Carl Sagan

CARL A. RUDISILL LIBRARY
LENOIR RHYNE COLLEGE

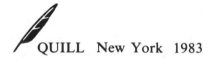

QUILL   New York   1983

523.109
F 4/r
130560
Jan.1985

Copyright © 1977. 1983 by Timothy Ferris

All rights reserved. No part of this book may be reproduced or utilized in any form or by any means, electronic or mechanical, including photocopying, recording or by any information storage and retrieval system, without permission in writing from the Publisher. Inquiries should be addressed to William Morrow and Company, Inc., 105 Madison Avenue, New York, N.Y. 10016.

Library of Congress Cataloging in Publication Data

Ferris, Timothy.
    The red limit.

    Bibliography: p.
    Includes index.
    1. Cosmology.   2. Expanding universe.   I. Title.
QB981.F36   1983   523.1'09   83-3068
ISBN 0-688-01836-X (pbk.)

Printed in the United States of America

First Quill Edition

1 2 3 4 5 6 7 8 9 10

BOOK DESIGN BY BERNARD SCHLEIFER

For JEAN BAIRD FERRIS
and in memory of
THOMAS ADDIS FERRIS *(1908–1977)*
and
BRUCE MACCANDLESS FERRIS *(1947–1968)*

PREFACE

We all dwell in a house of one room—the world with the firmament for its roof—and are sailing the celestial spaces without leaving any track.

—JOHN MUIR

The Red Limit endeavors to tell something of the story of how, in this century, we human beings first glimpsed the depths of the universe. I am grateful to its readers for the warm reception they afforded the book when it was originally published in 1977, and to the publishers for suggesting a revised edition. This edition has provided an opportunity to report on the latest developments in astronomy and observational cosmology, to add material more recently brought to light by historians of science, and to amplify or edit passages in the interest of enhanced thoroughness and clarity. The sections of photographs have been revised and enlarged as well.

Writing about events that have transpired recently in history presents the danger of distortion due to lack of perspective, but it also has its advantages, among them that many of the people involved are alive and can talk about what happened. This is particularly advantageous when researching the history of science, since openness and candor characterize science as a community. I am grateful to the astronomers, physicists and cosmologists who consented to be interviewed for this book, among them Ralph Alpher, Halton Arp, John Bahcall, William Baum, Geoffrey Burbidge, Margaret Burbidge, Robert

Dicke, Thomas Gold, Jesse Greenstein, Robert Herman, Tom Kinman, Frank Low, C. Roger Lynds, Arno Penzias, Allan Sandage, Maarten Schmidt, John Archibald Wheeler and Robert Wilson.

Research for the first edition was conducted principally at the libraries of Brooklyn College of the City University of New York, Hayden Planetarium, the California Institute of Technology, New York University, and at the public libraries of New York City, Miami, Florida, and Los Angeles, California, and was facilitated by the valuable help of many staff members there. Some of the quotations from John Wheeler appearing in Chapter 10 were taken from an interview conducted by Laurence B. Chase, who generously agreed to their reproduction here. David Allen of the Royal Greenwich Observatory kindly permitted me to borrow material from his accounts of life at the world's major observatories. My thanks to friends who offered encouragement and advice, among them Ken Broede, Nora Ephron, Karen Hitzig, Frederick Lunning, Thomas M. Powers, Paul Scanlon, Alex Shoumatoff, Erica Spellman, Jack Thibeau and, especially, Ann Druyan. The manuscript of the first edition was read closely by Bruce Partridge and Ralph Urban of Haverford College, who offered many valuable suggestions.

In preparing the revised edition I benefited from the efforts as well of John Bedke, Richard Dreiser, Rhea Goodwin, Douglas Kirkland, Jerome Kristian, Martha H. Liller, Dennis Meredith, Mary Lea Shane, Stephen Strom, Jurrie Van Der Woude, Margaret Weems, my colleagues at the University of Southern California, and by staff members too numerous to mention at the American Institute of Physics, the Australian National Observatory, California Institute of Technology, Cornell University,

Harvard University and the Harvard College Observatory, Kitt Peak National Observatory, Lick Observatory, the Mt. Wilson and Las Campanas Observatories, Princeton University, the University of California, the University of Colorado, the University of Texas, Yale University and Yerkes Observatory.

For her love, encouragement and advice, I am deeply indebted to Carolyn Zecca.

The first edition of *The Red Limit* was written in New York City between 1973 and 1977. The revised edition was completed in Los Angeles in 1983.

CONTENTS

INTRODUCTION

HUMAN BEINGS ARE, at least so far, an extraordinarily successful species, dominating the land, sea and air of their native planet, and now, in at least a preliminary way, setting forth to other places. The secret of our success is surely our curiosity, our intelligence, our manipulative abilities and our passion for exploration—qualities that have been extracted painfully through billions of years of biological evolution. It is in the nature of mankind and the corollary of our success to ask and answer questions, and the deeper the question the more characteristically human is the activity. Today we are finding that a host of issues, once the exclusive province of philosophy and theology, are slowly yielding to scientific inquiry, that most human of human inventions. The structure of matter, the nature of consciousness, the origin and fundamentals of life, the motions of continents, the intelligence of other animals, the possibility of life on other planets, the formation and destinies of worlds are all becoming accessible to the mind of man. This is not because we are today more intelligent than our ancestors of hundreds of thousands of years ago; but rather because our technology, the logical extension of our manipulative abilities, has at last reached a state of development adequate to probe these profound questions.

But the deepest of these questions is a subject thrilling even to contemplate: the origins, natures and fates not of worlds but of universes. As near as can be told, we have been preoccupied with this question as long as we have been human. It provides an essential part of the earliest body of myths and legends of cultures all over the planet Earth. Many of these legends are, while beautiful, simplistic to the point of explaining nothing, and the world is imagined created out of the void by a primal being or a primal couple whose origin remains uncontemplated. The Greeks called such a being Chaos. In African and Asian myths there is a cosmic egg from which the world forms. Such ideas are the product of an extrapolation from everyday human experience to the cosmos at large. But there is no guarantee that everyday experience—on an insignificant dust mote in an immensity of space, and in an instant encompassed by the course of ages—may have any bearing whatever on cosmological issues. The depth of our understanding and the quality of our insights have been powerfully prefigured by the environment in which we have evolved. When we move our attention to areas in which we have no previous evolutionary experience—quantum mechanics, the world of the very small, say, or black holes, the world of the very dense—we find that the universe is not in accord with commonsense ideas. The universe is, of course, not obliged to conform to everyday notions on a small and obscure planet. Still, some early cosmological ideas are marked by a more impressive subtlety, as, for example, the occasional myths which posit an infinite regression of causes, or the view, in Book 10 of the *Rig Veda,* that even the gods may be ignorant of the origin of the universe.

We humans are far from being gods. And yet, in the decades of the twentieth century, we have been privileged to pierce at least a few of the veils that have clouded this subject. We have developed powerful new ways of thinking about the universe, largely connected with Einstein's theory of general relativity; and powerful methods of viewing the universe, chiefly large optical and sensitive radio telescopes. All cosmological questions are by no means answered. But a picture is emerging which in scale, scope and subtlety is one of the great triumphs of the human mind and hand.

We have accumulated not speculation but reasonably hard evidence on the extent, shape, age, contents and ultimate fate of the entire universe, and speculation has now expanded to include other possible universes, constructed on different principles and obeying different physical laws from those which we find apply to ours. With the further application of existing groundbased optical and radio telescopes, and particularly with the future launchings of large space telescopes, it is possible that almost all of the basic cosmological questions will be answered.

The ones that may long elude us are in some sense ultimate questions: why the physical laws are the way they are, where the matter of the universe came from "in the first place," and what was there before the universe existed. But these questions may not be real questions, operationally defined. They may be only apparent questions,

tied to our use of words. If, for example, the universe is
infinitely old, no question arises about its origin. If we can
invent a source for the matter in the universe we immediately encounter the question of the origin of that source,
and thereby come face to face with the standard problem
of an infinite regression of causes. If we are trapped in a
universe with only one set of physical laws, it is difficult to
imagine experiments about which other categories of
physical laws are possible. Much has been learned about
cosmology recently; much will almost certainly be learned
in the coming decade or two. But there is some comfort in
the thought that we will never know everything. It would
be a very dull universe for any intelligent being were everything of importance to be known.

Because the pace of recent discovery has been so rapid
and the subject matter so removed from everyday experience, the excitement and exhilaration and passion of modern cosmology is not appreciated nearly as well as it
should be. There is a need for a comprehensible, accurate,
up-to-date discussion of cosmology which does not talk
down to the intelligent lay reader. I believe Timothy Ferris' *The Red Limit* is such a book. It is gracefully composed, studded with metaphors and similes of poetic
clarity and remarkably successful in conveying in words
some of the content and feel of the mathematics essential
to the subject. It has entrancing historical vignettes, many
forgotten or generally unknown. It does not shy away
from the vigorous controversy and strong personalities of
the leading cosmologists of our age. It does not evade philosophical implications. It is not bashful about pronounc-
ing mystery where we are still ignorant. For many readers
who have not previously encountered modern cosmological ideas, this book will provide a twofold revelation—

about the beauty and grandeur of the universe, and about the brilliance and tenacity of the human minds that occupy an obscure corner of that universe.

CARL SAGAN
*David Duncan Professor of Astronomy
   and Space Sciences
Director, Laboratory for Planetary
   Studies
Cornell University
Ithaca, New York*

When men lack a sence of awe, there will be disaster.
—LAO-TZU

THE
RED LIMIT

1

THE EXPANSION
OF THE UNIVERSE

*Although the myriad things are
many, their order is one.*
                    —CHUANG-TZU

IN THE TIME IT TAKES to read this sentence, the Earth will glide 200 miles in its orbit around the sun, the sun 3,000 miles in *its* orbit around the center of our galaxy, and 350,000 miles of additional space will have opened up between our galaxy and those of the Hydra cluster as the universe goes on expanding. The expansion of the universe is thought to have begun in a genesiac explosion (the "Big Bang") about 20 billion years ago. Astrophysicists and geologists estimate that the sun and its planets were born some 4.5 billion years ago, when the universe as we know it was something over two-thirds its present age. At a galactocentric velocity of 150 miles per second, the sun has wheeled around the center of the Milky Way Galaxy about twenty-five times since it was born. One percent of one orbit ago—since which time the sun has moved across the face of our galaxy by less an increment than the second hand of a clock consumes in one second—the species *Homo sapiens* evolved on the sun's third planet, Earth.

The Milky Way is one galaxy of perhaps two dozen that share membership in a cluster astronomers call the Local Group. Our nearest neighbors are two satellite galaxies, the Large and Small Magellanic Clouds, and a half dozen or so dwarfs. Two and a quarter million light-years away stands the Andromeda Galaxy, the dominant spiral of the group, an outsized replica of the Milky Way. The population of the Local Group is perhaps a thousand billion stars.

Beyond, billions more galaxies recede in profusion in deepening space. The antique light from many of them was near completion of its long journey to our eyes when life on Earth had evolved as high as the sea urchin.

The universe contains at least as many galaxies as there are stars in the Milky Way.

Look up. A moonless late summer evening is a good time in the northern hemisphere. Face due east and you are looking away from the plane of our galaxy; out there, past a few thousand foreground stars, lies intergalactic space. Seven degrees above the star Beta Andromeda, you should be able to make out the soft glow of the Andromeda Galaxy. Turn and face due west, near the bright navigational star Arcturus. Here you are looking pretty well out of the whole Local Group. The inky sky between the foreground stars harbors clouds of remote galaxies.

From northeast to southwest arcs the Milky Way, our galaxy viewed from within. Sweep it slowly with binoculars—the slower the better—from Cassiopeia in the northeast. You will see meadowlands of stars cut with hints of glowing gas. One huge cloud of dark dust and gas reveals itself as a rift dividing the Milky Way from Cygnus to the southern horizon, like a rip in the sky. Approaching Sagittarius in the south, your field of view becomes heaped with stars. You are looking toward the heart of our galaxy.

Our knowledge of the depth of the sky is new. Our ancestors tended to envision the sky as a domed roof; Lucretius was not alone in thinking it so low that a war cry might fetch an echo off it. To have discovered that the sky is instead bottomless, that it represents nothing less than a view of the universe as seen from within a major spiral galaxy, was a feat more prodigious, in terms of scale, than if, say, a band of protozoa in a Philippine tidal basin were to have charted the Pacific Ocean. The scientific discoveries required to begin mapping the universe in three dimensions came about in a revelatory flurry in the twentieth century but involved research dating back centuries earlier.

When Galileo turned his telescope on the Milky Way, he found that it was composed of millions of stars. This was the first evidence that stars might be distributed not at random but as part of a system—that, as we would say today, our sun and the stars we see in the sky are part of the disk of a spiral galaxy. Other galaxies were visible to Galileo, but they are so remote that their multitudes of stars blended together when viewed through his telescope. Consequently, Galileo was able to discern no fundamental difference between these "spiral nebulae" and the other nebulae that we today know are clouds of gas within our galaxy.

The eighteenth-century comet-hunter Charles Messier catalogued 103 nebulae, by way of warning other comet-seekers not to mistake them for legitimate prey. Many more were observed by the English astronomer William Herschel, who had a technologically premature passion for building big reflecting telescopes and who liked to boast, "I have looked farther into space than ever human being did before me." Herschel's son John continued his father's observations. Some of the nebulae looked like chalk-colored spiderweb tangled among the stars. Others were spiral in shape, resembling pinwheels. Throughout the nineteenth century, people were inclined to think that all the nebulae were gas or dust in our own stellar system. The one remarkable exception was the philosopher Immanuel Kant, who perceived, with little but reasoned intuition to guide him, that the delicate pinwheel nebulae might be galaxies.

In 1751 Kant, at that time a tutor in Königsberg, read a newspaper story about the speculative cosmologies of Thomas Wright, a pious English surveyor and amateur scientist who authored several theories of the cosmos. Some of Wright's models were mutually contradictory—he proposed variously that the Milky Way was spherical or flat like a grindstone, composed of stars like the sun or just an illusion—but the contradictions did not seem to bother him; he wrote his theories as kapellmeisters composed cantatas, as offerings to the greater glory of God. By virtue of a lucky accident that marks one of journalism's occasional contributions to science, the newspaper account Kant read badly oversimplified Wright, giving Kant the impression that Wright saw the Milky Way as a thin disk composed of stars, which he did not but which in fact it is. The idea appealed to Kant, and after four years of study he published, at age thirty-one, a slim, anonymous book that struck close to the truth. Titled *General History of Nature and Theory of the Heavens,* it proposed, correctly, that some nebulae, those clearly associated with stars, lie within our own Milky Way, while others, the spirals or oval-shaped nebulae, are separate Milky Ways at enormous distances. This made Kant the first to guess the true nature of the spiral "nebulae." The book attracted little attention, in part because there was then no way to test Kant's theory. To do so would require the spectroscope.

In 1802 the English physicist William Wollaston found that by placing a thin slit in front of a prism he could break sunlight down into component parts so sharply defined that dark lines appeared along the spectrum, signaling the absorption of characteristic frequencies of light by atoms of various elements in the sun's outer layers. The ability of a prism to dissect sunlight into colors had been known for a long time (Isaac Newton discovered that in 1666) but Wollaston's refinements turned spectroscopy from a diversion into a scientific tool. A Bavarian optician, Joseph von Fraunhofer, soon built a more advanced spectroscope and found that the sun's spectrum, from deep red to violet, was interrupted by hundreds of black lines that resembled the gaps between piano keys.

John Herschel had learned that each chemical element, when heated and its glow analyzed through one of the new laboratory spectroscopes, produced a characteristic spectrum all its own. The English chemist Robert Bunsen and the German physicist Gustav Kirchhoff in 1859 compared laboratory spectra with a solar spectrum and found telltale lines of hydrogen, iron, sodium, magnesium, nickel and calcium in the sun. A question once considered the epitome of the unknowable—what are stars made of—could now be answered, using the spectroscope.

The first astronomer to train a spectroscope on the stars was Sir William Huggins, a wealthy gentleman who maintained an observatory on the roof of his home in Tulse Hill, London. Educated as a chemist, Huggins built a spectroscope, attached it to his telescope and immersed himself in the most exciting work of his life. Each distant star obligingly revealed to his spectroscope the chemical elements it was made of. "Nearly every night's work was red-lettered by some discovery," Huggins wrote happily. After satiating himself on starlight, he turned in 1864 to the nebulae. His results helped support Kant's hypothesis. He found that the nebulae were divided into two kinds: Some were composed of gas, while others displayed spectra much like the sun's, suggesting that they were made up of stars. The spiral nebulae Huggins examined all had sunlike spectra.

Two theories prevailed concerning the spirals. Some held, with Kant, that they were external systems of stars. The majority leaned toward the view that they were whirlpools of gas, relatively nearby, each in the process of forming a new star. The whirlpool model had been put forth in 1796 by Pierre-Simon Laplace, a mathematician famous for his elegant analysis of how the planets move in their orbits. Laplace's book on the subject, *Essay on the System of the World,* sold well, while Kant's was almost unknown. And Laplace's model had a certain grace, even though he was misguided in applying it to the spiral nebulae. Huggins' spectroscope might have cleared up the question right away. Instead, the issue was confused by a cataclysm that had occurred two million years earlier.

In 2 million B.C., Earth time, a star near the center of the Andromeda Galaxy exploded with tremendous violence. Some stars dying in this fashion do so as novae, a few as the still more powerful supernovae; this was an especially violent supernova. In a matter of days the star disgorged itself of energy and collapsed into a cinder. Out through surrounding space spread a shell of light several light-days thick. Within sixty thousand years the light had cleared its native galaxy. In a hundred thousand years it swept through the two satellite galaxies, M32 and NGC 205, that ride like pilot fish near the big spiral. For two million more years, light from the supernovae sped across intergalactic space. It reached the outposts of our galaxy roughly when the pyramids of Egypt were being built. To the temporary misfortune of science, it intercepted Earth in 1885, just in time to confuse the debate over the nebulae. An astronomer named Ernst Hartwig noticed the new star on August 20.

The Laplace hypothesis seemed confirmed. Here was a

"new star," just where Laplace anticipated one, in the center of a swirling spiral nebula. The star soon faded, but that could perhaps be explained as its having settled down after a traumatic birth. Kant's correct perception that the Andromeda spiral was a galaxy seemed discredited.

The overriding problem in understanding nebulae and stars alike was that astronomers did not know the distances to most of the objects they observed. A jewellike star floating prettily in a telescope's field of view might be a hundred light-years away or a hundred thousand; so might the Andromeda spiral.

Distances to a few hundred nearby stars had been painstakingly worked out by the parallax method, which exploits the fact that the Earth's motion around the sun alters our perspective so that neighboring stars appear to shift slightly relative to those in the background. But this method worked well only up to a few tens of light-years away. Beyond that, astronomers knew where stars were in the sky but not where they were in space. Until the third dimension could be supplied, it was but conjecture to talk of whether there were such things as galaxies, and if so whether ours was shaped like a cake, a pinwheel or a smoke ring. Herschel, again ahead of his time, tried to construct a model of the Milky Way by counting stars in all directions and assuming them to be of about the same intrinsic brightness, but stars come in all sorts of intrinsic brightnesses, and the task defeated him. He ended up with a model of the Milky Way that resembled a thick botched pancake. He went on to other matters.

Stuck in two dimensions, astronomers in the closing decades of the nineteenth century shelved most big questions and bent themselves to less inspiring but more reliable work. They compiled star charts and assembled huge catalogues that listed the names, spectra and color of stars. Late-nineteenth-century university astronomy departments resembled factories of the period, where imperious bosses oversaw rooms full of employees whose repetitious toil threatened their eyesight. But these methods produced results. From the columns of dry figures came answers to secrets of the stars.

At Harvard College Observatory, the menial jobs were held mostly by women. Edward Charles Pickering hired them, gave them the title "computer" and paid them twenty-five cents an hour to fill blank catalogue pages with tiny black ink numbers, no mistakes permitted. A methodical man, Pickering came to Harvard from M.I.T. in 1876, stayed until his death in 1919 and spent much of that time photographing and indexing stars. With the help of his younger brother William he set up an observatory in the clear air of Flagstaff, Arizona, and another in Arequipa, Peru, to photograph the northern and southern skies systematically with standardized films and exposure times. Spectra for thousands of stars also were recorded at the two stations. The exposed plates were sent to Cambridge. Over the years Pickering's collection grew to a quarter of a million photographs, 225,000 spectra, and data filling seventy volumes of the Harvard Observatory *Annals*. About forty women worked as computers for Pickering. Most came and went silently, but a few persisted in trying to make some sense out of the forest of data they had been asked to harvest.

One was Annie Jump Cannon. Pickering befriended her and allowed her to sit in on his physics lectures at M.I.T., normally barred to women. While still an undergraduate she established a women's physics laboratory, then took a B.S. at Wellesley and went on to index Harvard sky plates for Pickering without pay. By 1896, at age thirty-three, she had become expert on the spectra of stars. Pickering gave her a small salary. She stayed at Harvard for forty-two years, discovering over three hundred variable stars and indexing three hundred thousand spectra. Fourteen years of this won her Pickering's personal accolade "astronomer," but, while she gained an international reputation and received honorary degrees from Oxford and three other universities, her name did not appear in the Harvard catalogue. Her presence was first officially acknowledged by the university in 1938, when she was seventy-five years old.

Another "computer," Henrietta Swan Leavitt, discovered what is called the period-luminosity relationship in Cepheid variable stars. Leavitt came to Harvard from the Society for the Intercollegiate Instruction of Women (later Radcliffe) in 1895. Pickering stacked plates from the Peru station in front of her and told her to look for variable stars. She would study two photographs taken of the same area of sky at two different dates and check whether any of the thousands of stars had changed in apparent brightness, betraying itself as variable.

Variable stars had been observed for centuries and were known to fall into two categories, the eclipsing binaries and what might be called genuine variables. Eclipsing binaries are systems of two or more stars that happen to be oriented in space so that one star, seen from our viewpoint, periodically passes in front of the other. True variables are stars that pulsate; their intrinsic brightness varies with time. A few take a year or more per pulse, but others, less leisurely, vary over periods of from a few hours to about a month. These more rapid variables included the Cepheids, so called because the first one identified was in the constellation Cepheus, named after the mythological king of Ethiopia who fathered Andromeda. The Cepheid variable stars proved to be the powerful distance indicators astronomers needed.

Henrietta Leavitt examined thousands of Cepheids on the Harvard plates—she discovered twenty-four hundred herself—and as time passed she began to perceive a pattern emerging. The brighter the Cepheid variable star, the longer it took to go through a cycle of brightness. She was able to make this important discovery because, as it happened, many of the Cepheids she was assigned to study were in the Large and Small Magellanic Clouds, two satellite galaxies of the Milky Way 150,000 light-years distant. Pickering and his staff of computers did not know what the Clouds were, but an important effect operated anyway: All the Cepheids in Clouds were bound up together at roughly the same distance, like fireflies in a bottle, so that the confusing differences in brightness caused by varying distance were suppressed and the inherent relationship of period to luminosity could reveal itself. If the true brightness, called absolute magnitude, of a single Cepheid star could now be determined, distances all over the cosmos might be measured.

Finding the distance to any single Cepheid proved difficult because none was nearby. But several astronomers, by exploiting a statistical method involving the sun's drift among its fellow stars, managed to get a fair estimate of the distance to a few short-period Cepheids. Now, it appeared, the genuine distance of any Cepheid variable could be determined from knowing only its apparent brightness and the period of its variability. Astronomy had found its third dimension.

Enter Harlow Shapley, a young man with slicked-back black hair, steady dark eyes and an appetite for fame. Born in Missouri, he became a newspaper reporter at age fifteen, banging out stories for the Chanute, Kansas, *Daily Sun,* then the Joplin, Missouri, *Times* (where an editor once sent him to debunk a horse that could count, not in order to print an exposé but to blackmail the carnival into advertising in the paper.) Shapley felt that formal training might bring him loftier assignments, and in 1907 he arrived at the University of Missouri to study journalism, only to find that the journalism school would not open for another year. Determined to study something, he chose astronomy. (Years later he insisted that he made his choice because astronomy, starting with "A," appeared near the front of the course listings.) He stayed at Missouri four years, won a fellowship and went to Princeton, where he found the observatory director, Henry Norris Russell, wrestling with the problem of eclipsing binary stars. These double-star systems were too distant to be resolved in a telescope. The fact that they were double could be detected only because their light output changed when one star eclipsed the other. The goal was to determine the full story of each double-star system—what it looked like, what it was made of, how it behaved. It was the sort of assignment a young astronomer could build a career on.

Shapley gave himself over to it. He pushed the Princeton equipment to its limits, then followed an elaborate line of reasoning to deduce portraits of the eclipsing star systems. Soon he was boasting that with only a telescope, spectroscope and photometer (this last instrument to measure brightness) he could tell twenty different things about an eclipsing binary star that looked to the eye like nothing more than a pinpoint of light. He could tell how many stars were involved in the system, how far apart they were in space, how fast they orbited each other and, very important, their distance from Earth. Shapley's groundbreaking work was rewarded with a job at Mt. Wilson Observatory, in California, home of the world's largest telescope, at a salary of $135 a month.

Mt. Wilson in 1914 was a rugged place, accessible only by a nine-mile path that climbed 6,000 feet up which everything on the mountain had been brought by backpack or burro. The nights were dark, the skies filled with steadily blazing stars. On the peak stood a telescope with a 60-inch mirror and the site where George Ellery Hale, the observatory director, planned to build a 100-inch mirror.

When Shapley arrived, he was prepared to turn his attention from binary stars to the Cepheid variables. He carried with him papers by his Princeton mentor Russell and by Russell's collaborator, the Danish researcher Ejnar Hertzsprung, who had been the first astronomer to employ Cepheids as indicators of distance. The papers estimated the intrinsic brightness of a few nearby Cepheids. These Cepheids were short-term variables compared with the ones Henrietta Leavitt had found in the Magellanic Clouds—they pulsed in periods of only a day or so, compared to weeks or more for hers—but it seemed reasonable to assume that all were varieties of the same sort of star. Shapley rented a house in Pasadena with his bride, Martha, and took pack trains up the mountain to chart the depths of the sky, using the Cepheid variables as his beacons. He felt, he later wrote, "confident that I could do something significant at Mount Wilson if the people there gave me a chance. . . . My desire, almost from the first, was to get distances."

Solomon Bailey of Harvard had advised young Shapley that the place to look for Cepheids was in globular clusters. The globulars, spectacular cities of stars, had long fascinated observers. Bailey had found that for some reason globular clusters were full of short-period Cepheid variables. Shapley investigated with the 60-inch telescope, measured the apparent brightness of the Cepheids and the time each took to vary. Then, by comparing the result with Russell and Hertzsprung's information on absolute Cepheid brightness, he estimated the distances of several globular clusters.

About one hundred globular clusters associated with our galaxy are visible from Earth. Shapley was able to locate Cepheid variable stars in a dozen of the closer ones. He then extended his distance measurements deeper into space, employing every technique he could borrow or invent to map the realm of the globular cluster in three dimensions. In each of the nearby clusters he isolated the most luminous stars—the red giants and supergiants—and systematically compared their apparent brightness with that of the Cepheids. When he had done this often enough he felt he had a pretty good idea of the inherent brightness (called absolute magnitude) of the giant stars. Then he abandoned the Cepheid variables and used still brighter giant stars as his beacons, or "standard candles," to get distances to more remote globular clusters where he could not identify Cepheids.

To envision how the standard candle method works, imagine that the globular star clusters are ships anchored at various distances across a dark and boundless sea. We are perched in a lighthouse and equipped with a telescope, spectroscope and photometer. Our assignment is to measure the distances of the ships. It is dark. All we can see through our telescope are the lights on each ship. Using the spectroscope, we determine the chemical composition of the element burning in each lamp. This tells us what sort of lights each ship has—incandescent pilot lights and running lights, a kerosene lantern in the wheelhouse, a fluorescent lamp in the galley and so forth. Using the photometer, we measure the relative brightness of the lamps on any given ship. The pilot light on a freighter, for instance, might be ten times as bright as the kerosene lantern in the wheelhouse. If this relationship between the brightness of pilot lights and kerosene lanterns seems to hold for each ship that has both, we can hypothesize that *all* pilot lights, generally speaking, are ten times brighter than kerosene lanterns, a hypothesis that enables us to estimate the distances of ships that have only one or the other kind of light. If, for instance, the pilot light aboard *this* ship appears to be ten times brighter than the kerosene lantern on *that* ship, we can assume that both ships are about equally distant.

But to estimate the absolute rather than the relative distances of the ships, we must ascertain the actual distance of at least one ship. Finally we find it—a little harbor scow so nearby that its position seems to shift against that of the background ships when we cross the catwalk on the lighthouse turret. By triangulation from either side of the catwalk—a procedure analogous to an astronomer's determination of the parallax of nearby stars—we estimate that the harbor scow is three miles away. We then compare the brightnesses of the scow's kerosene lamp to those on other ships that have similar lamps. We find that those ships are six, eight and ten miles away. At this distance the kerosene lamps are becoming too dim to see, but since many of the ships also have pilot lights and running lights, we can employ these brighter lights as standard candles. In this way, we extend our distance estimates ever farther across the sea. The method is fraught with the potential for error: Our calculations in the lighthouse will go astray if one ship's pilot light is intrinsically brighter than another's; and if the scow had an unusually bright or dim kerosene lantern, *all* our distance estimates will be off. But it's a start.

This is what Shapley did with globular clusters. And he employed other procedures as well, tried anything that would help him perceive the realm of the globulars in three dimensions.

His work came in a storm. Of twenty papers published in the 1918–19 issues of the Mt. Wilson *Contributions,* half were by Shapley. Combining mountains of fresh data from the 60-inch telescope with Shapley's bold, rigorous reasoning, they helped set a new tone in astronomical research.

Shapley's three-dimensional map of the world of globular clusters showed that they were arranged in a vast spherical volume of space, as if they themselves were members of a sort of super globular cluster. The center of this sphere was nowhere near the sun but lay instead tens of thousands of light-years away, in the direction of the constellation Sagittarius.

Shapley now made an intuitive leap. He asserted that the center of the realm of the globular clusters was the center of our stellar system, the Milky Way, as well.

Copernicus had maintained that Earth was not the center of the universe; Shapley maintained that the sun wasn't either. We live, he saw, in our galaxy's suburbs.

Though Shapley was right about our place in the Milky Way, he was wrong about its size. He estimated the diameter of our galaxy at 250,000 light-years, with the sun situated 50,000 light-years from its center. The correct values are more nearly 100,000 light-years diameter, with the sun 30,000 light-years out toward one edge. Shapley failed to appreciate that interstellar dust and gas dim our view across the galaxy, and that the obscuration is most pronounced along the galactic plane, which was just where most of his globular clusters lay. The effect was like that of fog upon the distance estimates of our ship-watcher in the lighthouse. Measuring a giant star in a remote globular, Shapley attributed its dimness to sheer distance, when in fact it was caused in part by absorption of light by interstellar dust and gas. As a result he tended to put the clusters, and consequently the center of the galaxy, too far away.

The mistake was not crippling so far as understanding our galaxy was concerned, but it got Shapley into trouble on the question of whether other galaxies exist. Some thought, quite correctly, that the Magellanic Clouds were satellites of the Milky Way. But Shapley's Milky Way was so large that it engulfed the Clouds. Similarly, it was hard to see how the spiral "nebulae" could be separate galaxies, unless ours was somehow a bloated monarch and all the others dwarfs. Shapley took an unyielding stand in defense of his picture of the Milky Way, inflated figures and all. He denied that the spiral nebulae were other galaxies. Since the spirals were the best candidates for the role of galaxies, he seemed to be implying that our Milky Way was all there was to the universe.

From Lick Observatory, up the California coast near San Jose, an astronomer named Heber Curtis began sniping at Shapley's conclusions. This was an early volley in a long war. Lick Observatory was associated with the University of California, Mt. Wilson with the Carnegie Institution and later Caltech, and the two were to quarrel for years. Lick astronomers chided their Mt. Wilson colleagues for building grandiose theories upon scant data. Mt. Wilson, in turn, viewed Lick as unimaginative and overcautious. As Shapley wrote Russell in 1920, "Lick believes practically nothing that you or I or Eddington or Hertzsprung have done in the way of interpretation. . . . They play safe. . . ."

Curtis thought Shapley's model of the Milky Way ludicrously outsized. An advocate of the "island universe" theory, he envisioned the spiral nebulae as galaxies much like ours. Shapley saw the universe as dominated by what he called our "Big Galaxy," with the spirals but clouds tangled in its outer branches. Curtis' attempts to demote the status of the Big Galaxy aroused in Shapley something of the dread felt by a patient facing the surgeon's scalpel; for Curtis to prevail, Shapley wrote Russell, he "must shrink my galactic system. . . ."

A debate between the two astronomers was arranged by the National Academy of Sciences in Washington, D.C. They rode the Southern Pacific to the East together, restricting themselves to nonscientific small talk so their debate might be fresh. It took place on April 26, 1920, with Albert Einstein in the audience.

During the debate and in the months of discussions that followed, Curtis focused on the weakest aspect of Shapley's model, his contention that the spiral nebulae are local. Shapley responded with two arguments. One was that exploding stars in the spiral nebulae, like the 1885 supernovae in the Andromeda Nebula, proved that the spirals could not be distant galaxies: For the supernovae to appear so bright at intergalactic distances they would have to shine with the light of hundreds of millions of ordinary stars, and this seemed an obvious absurdity. Shapley's other major argument derived from observations conducted by his friend and colleague Adrian van Maanen, indicating that the spiral nebulae were spinning rapidly—some so rapidly that, were they galaxy-sized, the outer parts of the disk would have to be moving at velocities greater than the speed of light, an impossibility.

Both these arguments proved to be erroneous. For one thing, supernovae *do* shine with extraordinary brilliance. For another, Van Maanen's observations were spurious. The stately spin of a spiral galaxy consumes hundreds of millions of years; it cannot be detected by comparing photographs taken a few years apart, Van Maanen's method. Yet Van Maanen could find no systematic error in his data even when, faced with a growing body of conflicting evidence, he reviewed his work to see how he might have erred. An honest and able scientist, he may have been misled into seeing what he expected to see. Whatever was responsible, his results eventually lost out against the verdict of the sky.

History's verdict in the Shapley-Curtis debate was mixed. Today we think that Shapley was right in proposing that the sun is located well away from the center of our galaxy, wrong in the values he assigned to its dimensions. Curtis was right in thinking of the spiral nebulae as galaxies, wrong in imagining that the sun sits at the center of our galaxy. Both men agreed that absorption of starlight by interstellar dust and gas is negligible; on that point, both were wrong.

The consensus at the time was that Curtis had prevailed. To his intense annoyance, Shapley found himself widely judged the loser in what ranked, after the Scopes trial, as the most widely publicized American scientific debate of the time. He never forgot it. Decades later he was still fuming at Curtis for, he felt, drawing him off the subject, and at Van Maanen for botching his observations of the spirals. In his memoirs he wrote, "They wonder why Shapley made this blunder. The reason he made it was that Van Maanen was his friend and he believed in friends!"

Shapley soon quit Mt. Wilson and accepted the directorship of Harvard College Observatory. (Pickering had died in 1919.) Friends urged him to reconsider; compared to Mt. Wilson, where the heavens were explored nightly with the best telescopes in the world, Harvard was a countinghouse. But Shapley went off to Cambridge. He became a distinguished figure, president of both the American Academy of Arts and Sciences and the American Association for the Advancement of Science, winner of a drawerful of medals, author of a half-dozen popular books on astronomy, a radio personality. He dabbled in politics, helped found UNESCO, was attacked by Senator Joseph McCarthy. When the spiral "nebulae" proved to be galaxies after all, the observatory graduated from cataloguing stars to cataloguing galaxies, by the thousands, with the help of new generations of women assistants for whose toil Shapley coined the term "girl hours."

While Shapley kept busy, breakthroughs were being made at Mt. Wilson. If this pained him he never said so, though he conceded, "sometimes I have thought that anything I did after that [Mt. Wilson] was anticlimactic." Back on the drafty peak of Mt. Wilson, Shapley learned, Edwin Hubble—Hubble of all people, cold and overbearing Hubble—was discovering the universe.

Shapley disliked Hubble. The man was impolite; when an article of Shapley's was sent him for checking, he scrawled "Of No Consequence" across the manuscript and returned it, and the comment was inadvertently set in type under Shapley's byline. He was ungenerous: "The work that Hubble did on galaxies was very largely using my methods. . . ." Shapley wrote. "He never acknowledged my priority, but there are people like that." Hubble spoke with an Oxford accent. This infuriated Shapley, who, like Hubble, came from Missouri. Shapley strongly suspected that if Hubble were awakened in the middle of the night he would talk like a Missourian.

A tall man with a strong jaw, thin mouth and a chilly gaze, Hubble did little to court the affection of his colleagues. A handful of friends found him warm and even charming, but to others he appeared arrogant and unprepossessing. A photograph taken of him dressed for trout fishing, flyrod in hand, staring into the camera, makes one feel for the trout. Most would admit, but few volunteer, that he was one of the greatest astronomers who ever lived.

Born in 1889 in Marshfield, Missouri, the fifth of seven children raised by a disciplinarian father, Hubble read Jules Verne and learned the constellations as a boy. He went to the University of Chicago on a scholarship and emerged as a quick student, a track and basketball letterman, and a skillful enough boxer to fight an exhibition with Georges Carpentier, the world light-heavyweight champion. Promoters looking for a "white hope" to fight Jack Johnson, the black heavyweight, asked Hubble to turn professional, but he chose instead to study law at Queens College, Oxford, as a Rhodes scholar. In 1913 he joined the Kentucky bar and practiced for a few months in Louisville, but the law bored him. "Astronomy mattered," he said.

Hubble re-enrolled at the University of Chicago, which operated Yerkes Observatory in Williams Bay, Wisconsin, near the Illinois border, home of the world's largest refracting telescope.* Built in 1897 with a donation from Charles Tyson Yerkes, the Chicago streetcar baron, this instrument, with its 40-inch lens, marked the climax of the age when astronomical telescopes resembled oversized spyglasses. Lilliputian astronomers toiled beneath its towering pier and 36-foot riveted steel tube; to reach the eyepiece they climbed bleacherlike wooden scaffolds that, with a little wrestling, could be moved around the observatory floor as the telescope tracked stars. Hubble, despite a few interruptions—he was stabbed in the back by a would-be thief in the woods one summer and once dove into Williams Bay to save a woman from drowning—got his Ph.D. in astronomy at Yerkes. In World War I, he

---

* The cardinal virtue of an astronomical telescope is to gather light. Refracting telescopes, from Galileo's through to the Yerkes refractor, do this by means of a lens. But the weight and thickness of the glass renders impractical refractors of more than about 40 inches diameter. Therefore the major telescopes of the twentieth century have been reflectors, using a curved mirror to gather the light and bring it to a focus. The mirror can be supported from the back for structural integrity, and less light is lost than with a refractor.

enlisted in the infantry and was wounded in France by shell fragments in the right arm. He arrived at Mt. Wilson in 1919, lured by the new 100-inch telescope.

From the outset Hubble's preoccupations were the spiral nebulae. He wrote his graduate thesis on the subject, arguing that astronomers ought to proceed on the assumption that the spirals were galaxies, since that line of inquiry, if justified, would lead to great things.*

Evidence that he was on the right track had come in July 1917, when Mt. Wilson's expert optician George Ritchey noticed a point of light on an old photograph of a spiral nebula in Cygnus and correctly surmised that it was a nova. Ritchey and Curtis then searched photos of the Andromeda Nebula in the files of both Mt. Wilson and Lick observatories and found two previously unnoticed novae in plates Ritchey himself had made. Luckily both appeared in sequences of photos made on more than one night, so the light curve of each exploding star—the way it brightened, then dimmed—could be reconstructed and compared with the behavior of novae that had been observed in our galaxy. Clearly there was a resemblance. The relative dimness of the newly discovered exploding stars constituted the best evidence yet that the Andromeda spiral must be very far away.

Hubble first studied nebulae he thought were probably local. Some of these were objects he had known since boy-

---

* Hubble may have borrowed this line of thought from the English astronomer Arthur Stanley Eddington, who wrote in a book published two years earlier, "If the spiral nebulae are within the stellar system, we have no notion of what their nature may be. That hypothesis leads us to a full stop. . . . If, however, it is assumed that these nebulae are external to the stellar system, that they are in fact systems coequal with our own, we have at least an hypothesis which can be followed up, and may throw some light on the problems that have been before us."

hood, like the beautiful Orion nebula. Others were dimmer. Using Mt. Wilson telescopes ranging from the little
10-inch Cooke camera up to the 60-inch reflector, after five
years he had sorted the nearby, or "galactic," nebulae into
distinct categories.

In a thirty-seven-page paper submitted in May 1922 to
Hale's *Astrophysical Journal,* Hubble first paid homage to
nebulae observers as far back as Herschel, then set forth
his own findings. "The nebulosity seems to consist of
clouds of matter, molecules, dust, or perhaps larger particles, not hot enough to be self-luminous, but visible because of light excited by or reflected from involved or
neighboring stars," he wrote. Some nebulae shine by the
reflected light of nearby stars. Others, those where the star
involved is hot enough, are bombarded by starlight so intense that their atoms emit light of their own, like those in
a fluorescent lamp. Hubble showed that when ordinary
stars were involved, the spectrum of the nebula looked
like that of a star; we were simply seeing reflected starlight. Where the stars were brighter, the nebula were "excited" (as the spectroscopists say) into producing spectra
of their own. The result was an emission spectrum, so
called because the nebula itself emits, rather than just reflects, light. This explained why nebulosity surrounding
the Pleiades star cluster displays a starlike spectrum, while
the Orion nebula, though it looks much the same to the
eye, has an emission spectrum.

Hubble showed that emission nebulae like that in
Orion were always associated with very hot stars. Since the
stars were certainly within our galaxy, these nebulae must
be too. Hubble therefore listed them as "galactic." The
spirals, the globular clusters, and some elliptical and irregular nebulae he classified "non-galactic."

Though he doubtless felt all but certain that his "non-galactic" nebulae were galaxies like ours and that in investigating them he was therefore embarking upon an exploration of unprecedented expanse, Hubble avoided saying so. He saw no reason to become embroiled in the Shapley-Curtis controversy. "There appears to be a fundamental distinction between galactic and non-galactic nebulae," he wrote, but quickly added, with a bow to Shapley's side, "This does not mean that the latter class must be considered as 'outside' our galaxy." Years later, when he was being acclaimed as the man who unveiled the galaxies and opened our eyes to the universe, Hubble still stuck to his old terminology. "The term *nebulae* offers the values of tradition," he wrote, "the term *galaxies,* the glamour of romance."

To resolve one of the spirals into stars would establish without much doubt that it was a galaxy. But this task appeared to be too much for the 60-inch telescope. A few of Ritchey's superb photographs did seem to reveal a scattering of individual stars in some spirals, but Ritchey himself cautioned that this evidence was tentative at best; the spirals might be nearby swirls of gas with a few new stars embedded in them, as Laplace had envisioned. When the 100-inch came into operation, Hubble made photos that came tantalizingly close to resolving the spirals into stars and revealing them as galaxies. Examining the spiral nebulae under a magnifying glass, Hubble sometimes felt he was seeing only a smooth wash of light. Other times, depending, he said, "on what I had for breakfast," he felt certain that the millions of tiny grains in the soupy imprint of the spirals were not a photographic effect or the product of his imagination, but stars, that he was staring at a vast continent of stars. Ultimately the 100-inch telescope would solve the riddle of the spirals. But for the present the question remained unanswered.

A few astronomers who tackled the problem of the distances of the spirals produced results that today look prophetic. Knut Lundmark, visiting from Sweden, declared while at Lick in 1921 that he had identified individual stars in the spiral M33; assuming them to be as luminous as the brightest known stars in our galaxy, Lundmark estimated the distance of M33 at 1 million light-years—short of modern estimates, which range from 2.5 million to 3.5 million light-years, but enormous by the standards of the day. In the following year, the prolific Estonian astronomer Ernst Öpik published a paper arguing that the Andromeda spiral was 1.5 million light-years away. This came remarkably close to the modern estimate of about 2.2 million light-years—closer, in fact, than Hubble was to come for another 20 years. If Lundmark and Öpik were right, the spirals lay beyond the thrall even of Shapley's Big Galaxy. But their estimates depended in part upon the assumption that the spiral nebulae were galaxies, and this remained a tautology so long as the status of the spirals was precisely the point under debate.

To verify to everyone's satisfaction that the spirals were galaxies, stars of the sort that were being used as distance indicators in the Milky Way would have to be identified in the spirals. Cepheid variable stars were the best candidates. Cepheids are bright enough to be seen across intergalactic distances. If the spirals really were galaxies, Cepheid variable stars ought to be visible in them. By measuring how long a Cepheid took to vary in brightness, one could estimate its intrinsic brightness, infer its (approximate!) distance and decide whether it, and the spiral to which it belonged, lay beyond the Milky Way.

Hubble accordingly searched for Cepheid variable stars in the spirals. Late in 1923 his efforts were rewarded. Triumphantly he scrawled the abbreviation "VAR!" for variable on the photographic plate. "You will be interested to hear," he wrote Shapley on February 19, 1924, with characteristic dryness, "that I have found a Cepheid variable in the Andromeda Nebula (M31). I have followed the Nebula this season as closely as the weather permitted and in the last five months have netted nine novae and two variables." Shapley replied to Hubble that his letter was "the most entertaining piece of literature I have seen for a long time." His response seems almost restrained, considering how he must have felt. After all, Shapley had been a leader in exploiting the method of determining distances by observing Cepheid variable stars. Now Hubble had turned that very method into a tool for dismantling Shapley's Big Galaxy. "I do not know whether I am sorry or glad to see this break in the nebular problem. Perhaps both," Shapley wrote Hubble five months later, as the evidence mounted that the spiral nebulae were galaxies comparable to our own.

Hubble's paper announcing his discovery of Cepheid beacons in the Andromeda spiral was read in his absence, though in the presence of both Shapley and Curtis, when the American Astronomical Society convened in Washington, D.C., at the end of the year. "When Hubble's paper had been read," writes Allan Sandage in his book *The Hubble Atlas of Galaxies,* relying upon the recollections of his colleague Joel Stebbins, "the entire Society knew that the debate had come to an end, that the island-universe concept of the distribution of matter in space had been proved, and that an era of enlightenment in cosmology had begun."

Hubble, continuing to seek out variable stars in other galaxies, examined an odd clutch of stars labeled NGC 6822. This cluttered stellar colony had first been noted in 1886 by Edward E. Bernard, a self-educated American astronomer, then studied more closely by Charles Perrine, a British veteran of Lick who settled in Argentina to view the southern skies. Located near the rich star clouds of Sagittarius, NGC 6822 might have been mistaken for part of the Milky Way, but Perrine, an astute observer, saw that it was a distinct entity, inhabited by a variety of indigenous stars. If this were so, Hubble realized, it might be a galaxy close enough for ready examination. Over a period of two years he photographed NGC 6822 and found fifty Cepheids there. Their brightness, analyzed largely with Shapley's methods, put them several hundred thousand light-years away. Here clearly was an object independent of our galaxy.

NGC 6822 seemed so staggeringly distant that Hubble was relieved to find identifiable stars there; he had not been certain other galaxies would even obey the same physical laws as ours. Finding the Cepheids reassured Hubble that observing deep into the cosmos, while difficult, was not going to be impossible. "The principle of the uniformity of nature . . ." he wrote, "seems to rule undisturbed in this remote region of space."

Having tested his skills in the cosmic archipelago, Hubble steered for deeper water. For several years he had investigated the two largest spiral nebulae visible in Earth's skies. These hundreds of observations now generated two long papers, each a legion of research, bristling with data, one on the nebula M33 in Triangulum, in February 1926, the other on the Andromeda Nebula, in December 1928.

M33 is a lovely spiral, pretty as an ocean sunfish, oriented in space so we view it almost flat-on. Using a new, more sensitive photo emulsion, Hubble photographed it repeatedly with the 100-inch telescope on the clearest nights and finally succeeded in resolving it indisputably into stars. He then identified thirty-five of the stars as Cepheid variables (only three had been located before, and those tentatively) and from them obtained a distance estimate that put M33 even farther away than NGC 6822. Here was another galaxy, separate from and outside the Milky Way.

Hubble's Andromeda Nebula paper virtually broke the back of the subject. Previous astronomers had studied dozens of photographs of this entrancing spiral. Hubble scrutinized 350, of which he took more than 200 at the telescope himself. Two novae had been found there, by Ritchey; Hubble discovered 63, and reached the remarkable conclusion that the Andromeda is so populous a galaxy that 30 of its stars explode every year. He estimated its size and mass, and guessed correctly, though perhaps for the wrong reasons, that it might prove to be still larger.

With the publication of these papers the peoples of our planet discovered what, presumably, most cognizant beings evolving beneath clear skies learn sooner or later—that the cosmos is subdivided into galaxies.

That discovery, a long time in coming, was followed immediately by the discovery of the expansion of the universe.

Hubble had evidence of expansion on his desk while still proofreading his Andromeda Nebula paper. In the course of establishing the approximate distance, size and brightness of dozens of galaxies, he had also measured the velocity of each. His original purpose was to find out how rapidly the sun was moving as it took part in the wheeling of our galaxy; like passengers on a carousel, we ought to be able to determine how fast we are going by looking at objects in the distance. Hubble assumed that other galaxies would provide such points of reference, whether they were fixed or drifted at random in space. The sun's motion in our galaxy could be deduced from the apparent motion of the other galaxies, some of which presumably would appear to be approaching, others receding. Hubble did find traces of this effect, but overshadowing it he discovered something unexpected, monolithic and strange. Only a few local galaxies showed the sort of random drift anticipated. All the others appeared to be moving away. They were receding at remarkably high speeds, and, most startling, the farther away they were the faster they were receding.

There are two uncomplicated ways to explain this state of affairs. One is to say that our galaxy sits at the center of the cosmos and all the others are, for some reason, rushing away from us. This seems unlikely. The other explanation is that the universe is expanding. In any uniformly expanding entity—a spotted balloon, a loaf of rising raisin bread or a cosmos of galaxies—the distance between two given points will increase with time, and the greater the distance to start with, the faster it will increase. Arthur Stanley Eddington used to illustrate this effect by asking his students to envision what would happen if the lecture hall they were sitting in (the universe) were to double in size, carrying their desks and chairs (the galaxies) with it. Each would notice the nearby students had to move their desks only a few feet in order to double the distance, while those across the room had to move much farther; in any given period of time, therefore, the farther one looked across the hall, the faster one saw desks and chairs receding.

Hubble estimated the velocities of galaxies by breaking down their light with a spectrometer and measuring the degree to which lines in their spectra were offset. If a galaxy were approaching, its light would be shifted toward the short-wavelength end of the spectrum, the blue end. If it were receding, the light waves would be stretched out, shifting the lines toward the longer-wavelength red end. In all but the most nearby galaxies Hubble observed, the spectral lines were displaced toward the red. The more distant a galaxy, the greater the red shift of its light.

The historical scaffolding of Hubble's discovery dated back to the early nineteenth century, when Christian Doppler, a physicist teaching in Vienna, found that the wavelength of lines in the spectrum of a light source ought to shift if the source were moving rapidly toward or away from the observer, just as the pitch of an automobile horn sounds higher—shorter wavelength—if the car is approaching and lower—longer wavelength—if it is speeding away. The spectrum is said to be "Doppler shifted" by velocity. Spectral lines could be measured with considerable accuracy, so this meant the velocities of remote, bright objects, namely stars, could be too. James Keeler at Lick Observatory soon found a Doppler shift, in the spectrum of the bright star Arcturus—a blue shift, indicating that Arcturus and the sun were growing closer in the course of their common sweep in our galaxy. Thereafter Doppler shifts were widely employed in finding the velocities of stars.

The first sustained attempt to measure Doppler shifts in the light of galaxies came about prior to Hubble, as an inspiration of the eccentric astronomer Percival Lowell. Lowell came from a prominent Boston family—his great-

great-grandfather John was a member of the Continental Congress, his brother Abbott was president of Harvard, his sister Army was the poet and biographer of Keats—and he combined a passion for astronomy with the money to do something about it. On a pine-shrouded peak outside Flagstaff, Arizona, he built Lowell Observatory and there embarked on a scientific career informed by motives as much poetic as scientific. Lowell was convinced that the canals of Mars, since found to be an optical illusion, were waterways constructed by a Martian civilization perishing of thirst. And he insisted, just when most professional astronomers were discarding the idea, that the spiral nebulae were solar systems in the making. This devotion of his to an outmoded theory of the nebulae led circuitously to important work presaging Hubble's.

Lowell hired, perhaps as a counterbalance to his own flamboyance, a careful observer named Vesto Slipher. Later he made him observatory director. Patient and fastidious as Lowell was impulsive, Slipher never cut corners, never published a result until he felt sure it was correct. He wore a suit and kept his tie knotted even when alone at the telescope at night. Lowell encouraged him to make spectra of the spirals and to look for Doppler shifts, expecting that the results would support the old Laplace theory. Slipher himself felt this was unlikely. He thought the spirals were going to turn out to be galaxies. But he dutifully set to work taking spectra of them and measuring Doppler shifts, repeating difficult observations many times to insure accuracy. He had the field virtually to himself for more than a decade; Lowell was involved in it for the wrong reasons, and nobody else wanted to get into it much at all. By 1925 he had obtained Doppler shifts for forty-five spirals.

Slipher was surprised to find that nearly all the spiral nebulae were receding. Their red shifts indicated that some were hurtling away at as much as a thousand miles per second. This was strong evidence against the Laplacian hypothesis favored by Slipher's employer: Nebulae rushing into intergalactic space could hardly transform themselves into stars that lingered here in the Milky Way. But in itself, Slipher's work did not reveal that the universe is expanding. A few astronomers who saw his data, notably Carl Wirtz in Germany, thought they perceived traces of a correlation between the distances of the nebulae and the velocities of recession, but to establish this astonishing relationship with any real conviction required persuasive estimates of the distances of the nebulae.

The man who got those distances was Hubble. In 1927 the sun's velocity of rotation in our own galaxy was finally measured fairly accurately, and it became possible for Hubble to establish that the galaxies behaved according to a grand and startling design—the farther away, the greater their red shift.

Characteristically, Hubble handled his discovery with long tongs. Word of what one astronomer later called "the most astounding fact of the twentieth century" came in a modest little paper that contained no mention of the expansion of the universe or, for that matter, of the universe. Titled "A Relation Between Distance and Radial Velocity Among Extra-Galactic Nebulae," it approached the subject via its original and most mundane application, as a complication in the problem of fixing more accurately the sun's velocity in our galaxy. "New data to be expected in the near future may modify the significance of the present investigation or, if confirmatory, will lead to a solution having many times the weight." That was as evocative as

Hubble got, in announcing that the universe, long assumed to reside in stately lassitude, was instead expanding from genesis toward destiny.

For all his caution, Hubble saw from the outset that the velocity-distance relation, if it held up, afforded him a powerful new way to chart deep space. The Cepheid variable stars had yielded distances to several galaxies. By studying bright giant stars in those galaxies (as Shapley had done with the globular clusters), Hubble was able to measure the distances of systems where the Cepheids were too faint to be seen. Beyond that, he had obtained some distances by using the total brightness of whole galaxies as his standard, but this method was very rough. Needed was a procedure for estimating the distances of still more remote galaxies. The red-shift-distance relationship promised to be just that. If a galaxy's velocity of recession was really an index to its distance, then the distances of galaxies far across the universe could be inferred just by measuring their red shifts.

Increasingly from 1928 on, Hubble entrusted the night-by-night task of exploring these uncharted reaches to Milton Humason, an unusual man. Humason showed up at Mt. Wilson while traveling around California in search of work. He was hired first as a mule packer, then as busboy in the mountaintop dining room and as observatory janitor. He asked a lot of questions. Shapley and some of the other astronomers, impressed by his curiosity and intelligence, had him promoted to night assistant on one of the smaller telescopes. Humason did the job with energy and wit, and he managed to wangle a few routine observing assignments of his own on the little 6-inch telescope, then the 10-inch. Hubble finally tried him out on the 100-inch and was pleased with the results. Because

Humason had only an eighth-grade education, Hale long denied him another promotion, but he finally capitulated and made him a staff member. Easygoing and unaffected, Humason played poker with the night assistants and joked with the visiting Einstein in the same unassuming manner. He became a liaison between the olympian Hubble and the other astronomers and served as Hubble's assistant at the telescope, shouldering the bulk of the observing while Hubble smoked his pipe and pored over the results. First Humason installed a specially designed "fast" lens that reduced the amount of time required to photograph dim objects with the 100-inch. After testing it on previously observed galaxies he moved into deeper space, where entire clusters of galaxies were, by Hubble's plan, to be assayed like grains in a prospector's pan. The telescope was trained on the Pegasus cluster, the Perseus, Coma and Boötes clusters of galaxies; and a science that a few years before had not been certain whether the cosmos ended with the Milky Way now charted territory hundreds of millions of light-years out.

On clear, moonless nights, Humason perched on a little steel platform 50 feet above the observatory floor, the black skeleton of the telescope tube silhouetted against the dome slit, stars and sky. A night assistant, holding Humason's old job, sat at the console, his face illuminated dimly by red dark-vision lamps. They worked to a continuo of mountain winds and the heavy ticking of a weight-driven brass clock that powered the telescope. Intermittent chimes signaled that Humason, up in the dark, had pressed a button altering the drive rate to keep his guide star in place. If the correction mechanism got balky, as it often did, Humason would hold the image in place by leaning on the massive telescope, sometimes climbing aboard it; he would perform whatever acrobatics were required to hold the field steady while the photographic emulsion steeped in the ancient light of galaxies.

These nights tested Humason physically as well as intellectually. An observatory dome cannot be sealed from the elements, because glass windows would steal precious light; it cannot be heated because the heated air would rise, setting up air currents, and the slightest air turbulence scrambles the image in a large telescope. In winter the observer huddles at the eyepiece, tears streaming down his cheeks, breathing deeply to ward off attacks of shivering, staring at a guide star which he cannot ignore for a moment without risking his time-exposure photograph or spectrum. As the telescope tracks across the sky the eyepiece can roll to diabolical orientations, prompting recollection of a Gilbert and Sullivan-style tune Shapley and his Harvard students composed to describe the ideal astronomer:

> His knee should bend and his neck should curl,
> His back should twist and his face should scowl,
> One eye should squint and the other protrude,
> And this should be his customary attitude.

The astronomer Rudolph Minkowski once endured five successive nights of this punishment to expose a single plate in an effort to record a very faint galaxy; then at dawn after the fifth night, exhausted, he went down to the observatory darkroom to develop it, where by mistake he immersed it not in developer but in fixer, destroying the image. He stayed in astronomy anyway.

By the time Humason reached the Boötes and Ursa Major clusters, he was dealing with galaxies too faint to be seen by eye even through the 100-inch telescope. His technique was to locate them in wide-field time-exposure photographs, then aim the big telescope by setting it on the proper celestial coordinates, hoping the galaxies were there in the black field. The setting circles employed for this were enshrined high up in the telescope mount; the night assistant read them from the console through two small telescopes of his own. Humason had to wait until the plate was developed before knowing if he had caught anything. In this fashion, he found galaxies receding at twenty-six thousand miles per second, one-seventh the velocity of light.

New, faster photo emulsions made it possible to record still more distant galaxies in the same exposure time. Now Humason's plates showed dense clouds of galaxies. An informal goal of Hubble's was to make a photograph that recorded as many galaxies as foreground stars. This was achieved on March 8, 1934. The print looked like beach sand scattered on a sheet of paper. About half the dots were stars in our galaxy; the others were galaxies, each home to billions of stars of its own. Hubble called it "a spectacular description of the penetrating power of the telescope." If he felt anything more, he said nothing about it.

2

THE UNIVERSE OF THE MIND: COSMOLOGY

It is obvious that we must regard the universe as extending infinitely, forever,
in every direction; or that we must regard it as not so extending. Both possibilities go beyond us.

—EDITORIAL IN
Scientific America,
March 13,1921

STUDENTS OF THE SKY from Lucretius to Newton who believed the cosmos to be infinite appear to have held that opinion less because they were enthusiastic about it than because the alternative seemed worse. A finite, three-dimensional universe would necessarily come to an end somewhere, an unimaginable situation. A conversant of the Chinese philosopher Wang Ch'ung two thousand years ago put the question in much the same words still heard today: "If Heaven has a boundary, what things could be outside it?" Lucretius regarded the notion of a finite universe as ridiculous. "Let us assume for the moment that the universe is limited," he wrote. "If a man advances so that he is at the very edge of the extreme boundary and hurls a swift spear, do you prefer that this spear, hurled with great force, go whither it was sent and fly far, or do you think that something can stop it and stand in its way?"

But the idea that the universe is infinite seems no less paradoxical. For one thing, Newtonian gravitation could not operate in an infinite universe full of stars and galaxies: The gravitational force exerted by the infinite numbers of distant stars would overwhelm local gravity at every point in the cosmos, tearing the stars and planets apart. (This point, which Newton overlooked, was discerned by two modern mathematicians, Carl Neumann and Hugo Seeliger.) Another less precise objection is that the concept of an infinite quantity of any *thing* raises intuitive problems. Alexander the Great purportedly wept upon being told infinite worlds exist because, he complained, "there is such a vast multitude of them," of which he had conquered only one. He might have consoled himself with the thought that a cosmos of infinite worlds might contain infinite numbers of Alexanders, busy conquering infinite Earths. Still, all their triumphs would amount to only an infinitesimal fraction of the whole, so perhaps our local Alexander sobbed justifiably. There is something unreasonable about the whole line of thought. If, instead, we propose that only space is infinite—that the realm of galaxies and stars ends somewhere, beyond which point ranges empty space—we find it difficult to explain what is meant by "space" in the reaches infinitely beyond matter. Space in the absence of objects is, if you think about it, indistinguishable from no space at all, a conclusion reached by Newton, Aristotle and Berkeley. Infinite and finite universes trouble the mind equally.

This ancient problem may have been solved in our century. One should be suspicious of claims that tough old questions recently have been answered. Still, some new things do appear under the sun, and the concept of a finite but unbounded universe was one. In a series of developments beginning in the nineteenth century and reaching fruition with the theory of general relativity in the twentieth, the finite but unbounded model released cosmology, the discipline concerned with the nature of the physical universe as a whole, from the finite/infinite dilemma. Two broad currents of thought led to it: the scientific positivism of Ernst Mach and the noneuclidean geometries of Karl Gauss, János Bolyai and Nikolai Lobachevski. They converged in Einstein.

First Mach. He was born in Turas, Moravia, in 1838. A poor student, he struggled a few years at an Austrian Ben-

William Herschel, an eighteenth-century composer and conductor of classical music who taught himself astronomy, built the largest working telescopes of his day. "I have looked farther into space," he boasted, "than ever human being did before me." *Photo: Yerkes Observatory*

George Ellery Hale built what were in succession the world's four largest telescopes, beginning with a refracting telescope with a lens 40 inches in diameter. Prior to installation at Yerkes Observatory in Wisconsin, it was displayed *(opposite)* at the 1893 World's Fair. *Photos: Yerkes Observatory*

Atop Mt. Wilson overlooking Los Angeles, Hale built an observatory equipped with the large telescopes that were to be instrumental in the discovery that we inhabit a galaxy in a universe of galaxies. *Clockwise from immediately above:* The peak, high above the clouds; the 60-inch telescope; construction of the dome for the 100-inch; assembly of its mounting; the completed 100-inch telescope, today a national engineering landmark and still in active use. *Photos: Carnegie Institution of Washington, Mt. Wilson and Las Campanas Observatories*

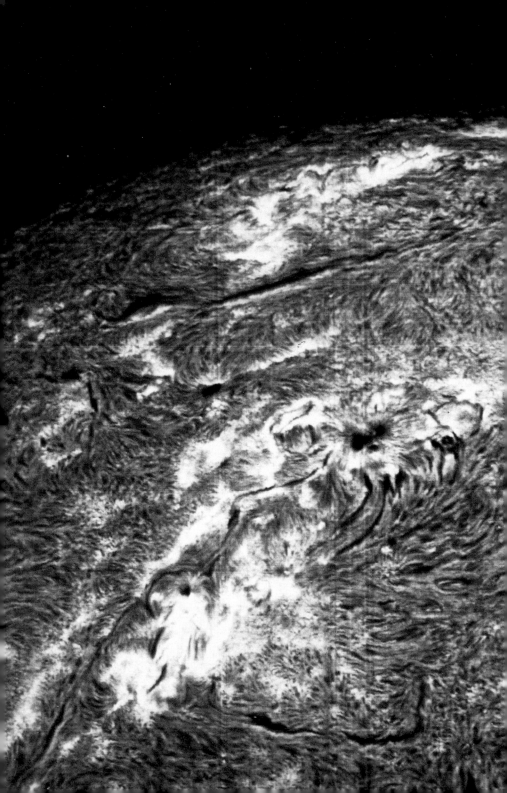

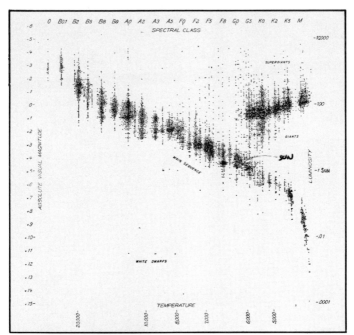

"The sun is a star, the only star whose phenomena can be studied in detail," wrote Hale, and he invented the spectroheliograph to permit detailed examination of the sun. *Opposite:* A spectroheliogram of the sun in the wavelengths of hydrogen light. *Above:* Ejnar Hertzsprung *(right)* and Henry Norris Russell independently discovered an elegant pattern of order underlying the diversity of the stars. When stars are plotted by color against true brightness, they learned, most stars fall along a path they called the "main sequence." The Hertzsprung-Russell diagram *(above, top)* is a sort of stellar Tree of Life, upon which may be traced the careers of stars. *Photos: Carnegie Institution of Washington, Mt. Wilson and Las Campanas Observatories; American Institute of Physics (2); Yerkes Observatory*

Albert Einstein, seen here with his friend Charlie Chaplin at the world premiere of *City Lights* in 1931, was inspired in composing his general theory of relativity by Ernst Mach *(opposite)*. Clearheaded and ceaselessly skeptical, Mach had impressed the young Einstein with his firm rejection of what he called the "monstrous conceptions of absolute space and absolute time" that ruled Newtonian physics. *Photos: Oxford University Press; American Institute of Physics*

Alexander Friedmann *(above)* found that according to Einstein's general theory of relativity, the universe cannot be static but must be either expanding or contracting. Einstein was sufficiently distressed that he marred the relativity equations by introducing a new term to make its theoretical universe stand still. Seven years later, astronomers independently discovered that the universe apparently *is* expanding, just as the theory implied. Arthur Stanley Eddington *(opposite)* helped introduce astronomers to the cosmological implications of Einstein's general relativity, and led a 1919 solar eclipse expedition that verified Einstein's prediction of the curvature of space. A visionary philosopher as well as astronomer and astrophysicist, Eddington wrote, "I only want to make vivid the wide interrelatedness of things." *Photos: American Institute of Physics; Yerkes Observatory*

Three of the astronomers who charted our galaxy appear in the photograph above. Jacobus Kapteyn *(third from the left)* determined that the stars we see in the sky are drifting along together as they take part in the wheeling of the Milky Way Galaxy. Karl Schwarzschild *(in white suit and hat)* brought relativity and quantum physics to bear upon the study of stars, playing a major role in extending the domain of physics to realms beyond the earth. Vesto Slipher *(top, far right)* discovered that the "spiral nebulae" (that is, galaxies) are hurtling away at high velocities. This set the stage for Edwin Hubble's discovery of the expansion of the universe. Percival Lowell *(opposite)*, champion of the illusory canals of Mars, had a way of doing the right work for the wrong reasons. He assigned Lowell Observatory staff astronomer Vesto Slipher to study motions in the spiral nebulae, thinking they were stars being born. Slipher instead found that the spirals are rushing away at enormous speed. This was the first observational evidence that we inhabit an expanding universe. *Photos: Yerkes Observatory*

Women hired as "computers" by the Harvard College Observatory painstakingly examined photographs of millions of stars, ultimately identifying variable stars that could be used to measure distances deep into space. Annie Jump Cannon *(seated, center)* located over three hundred variable stars. Denied a faculty post by Harvard, she was given an honorary degree by Oxford *(overleaf)*. Henrietta Swan Leavitt *(right of center, examining a plate)* discovered Cepheid variable stars, supergiants so bright that they can be seen in other galaxies. Indeed, it was in external galaxies—the Magellanic Clouds *(opposite)*—that she found the Cepheids, providing astronomy with a powerful new way to chart the depths of space. *Photos: Harvard College Observatory; except large Magellanic Cloud, Kitt Peak National Observatory*

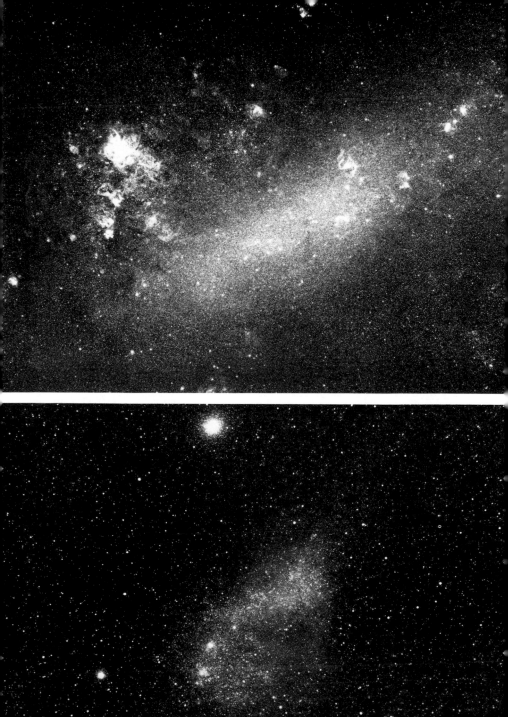

edictine academy, withdrew, served as a cabinetmaker's apprentice, returned to school at a Moravian gymnasium where he hated the classes but endured and finally gained admittance to college. Mach "looked into the world with the curious eyes of a child," Einstein wrote; a sense of the relativity of things seemed wired into his perceptions. Looking down from a bridge at a flowing stream, he became giddy with the feeling that the bridge was moving. A steamboat trip on the Elbe made him dizzy: He felt the land, not the boat, was gliding along. He frowned at the shabby dress of a man he saw on a train, then realized it was his own image in a mirror.

These uncensored perceptions, discomfiting in adolescence, served Mach well as an adult. He did distinguished work as a psychologist of human perception, as a physicist (the speed of aircraft relative to that of sound is measured in Mach numbers) and as a philosopher who insisted that scientists stick to data and shun assertions about the absolute reality, or lack of it, in their theories. Scientific "laws," Mach argued, result from the common experience of human beings and deserve promotion to no higher office. "Physics is not the whole universe. . . ." he wrote. "Our physical concepts, however close they come to the facts, must not be regarded as a complete and final expression of these facts." The "self-evident" postulates of Euclid and the "absolute" space of Newton, Mach felt, should be treated as no more than useful tools, to be discarded without remorse if tools better fitting the data were found.

In this, Mach was influenced by the eighteenth-century empiricist George Berkeley, who felt that the essential quality of the existence of a thing was our perceiving it, and so had little use for absolute space. In Berkeley's *Principles of Human Knowledge,* Mach read these words: "Absolute space, distinct from that which is perceived of sense and related bodies . . . cannot exist without the mind. . . . We cannot even frame an idea of *pure space* exclusive of all body." Imagine that only a single sphere exists alone in the universe, Berkeley wrote. We cannot envision that sphere revolving, because we have no reference points, nothing beyond the sphere to reveal that it *is* revolving. Now imagine *two* globes in an otherwise empty cosmos. Can they whirl round one another, like the rubber tips of a spinning baton? Again the idea is meaningless without some exterior point of reference. Newton's absolute space lends little help to our mind's eye. "But suppose that the sky of the fixed stars is created," Berkeley wrote. "Suddenly from the conception of the approach of the globes to different parts of that sky, the motion will be conceived."

At the time, Berkeley's argument had been lost in Newton's shadow. But Mach, reading it a century and a half later, was fascinated. He thought about an argument of Newton's concerning centripetal force. Hang a bucket of water by a rope and spin it; the water climbs up the inside of the bucket while depressing somewhat in the center. This behavior, Newton maintained, results from the water's motion relative to absolute space. The more Mach thought about this position of Newton's, the less tenable it appeared. What reality could be accorded to invisible, intangible space such that it could dictate terms to a palpable bucket of water? Carrying Berkeley a step further, Mach reasoned that if the bucket could be said to be spinning only relative to some exterior point of reference, then the *force* that made the water in the bucket climb its walls required those reference points too. "I have no objection to calling [it] rotation," Mach wrote, "so long as it be remembered that nothing is meant except relative rotation with respect to the fixed stars." He had no objection to talking about "centripetal force," so long as it was kept in mind that without the distant stars, no such force could be said to exist.

Mach's seemingly abstract point contained the seeds of a picture of the cosmos in which objects, instead of pushing and pulling one another by force, engage in a sort of silent collaboration that reads out to us as "laws" of nature. Inertia, the fundamental quality of matter from the standpoint of traditional science, was for Mach not inherent in each object but instead reflected its relationship with all the other objects in the universe. How hard you must push a boulder to get it moving, the force a batter needs to hit a home run, whether a coin stops rolling before falling down a grate were for Mach matters bartered among the stars. Mach did not explain, nor has anyone since, how his cosmic interaction actually worked; it has never been mathematically formulated. But Einstein was inspired by it. He called it "Mach's principle" and attempted to incorporate it into his then embryonic general theory of relativity.

Einstein visited Mach one day in autumn 1913, when Mach was seventy-five years old, semiparalyzed and sufficiently hardened in his views that he denied even the existence of atoms. Atomic theory was revolutionizing physics at the time, but Mach regarded the "atom" as just another concoction of the mind, useful as a conceptual tool but lacking any supportable claim to independent reality. Three years earlier, Mach had written bitterly to Max Planck, "If belief in the reality of atoms is so important to you, I cut myself off from the physicist's mode of thinking. . . . I decline with thanks the communion of the faithful. I prefer freedom of thought." Cut off he was. He received few visitors, and when they appeared in his small apartment in a Vienna suburb he shouted, "Please speak loudly to me. In addition to my other unpleasant characteristics I am also almost stone deaf." Einstein sat and shouted pleasantly about atoms.

There is no record whether the two talked about Mach's views of inertia, but the matter had captivated Einstein since his student days at the Federal Institute of Technology in Zurich. Just before visiting Mach, Einstein hinted in a speech that he was at work on a theory of gravitation that would be as far removed from Newton, he said, as radio was from Ben Franklin's kite and key.

Other antecedents of general relativity theory were the noneuclidean geometries. They originated in the failure of a hundred generations of mathematicians to derive Euclid's axiom of parallel lines—the axiom that can be stated, "Parallel lines never meet." Euclid, writing in the fourth century B.C., made no attempt to prove this assertion—as an axiom it was assumed to be self-evident—but it stood in curious contrast to the four axioms that precede it in his *Elements*. They are pithy; it is relatively lengthy and indirect. And while Euclid, having set forth the other axioms, proceeded to put them to work building his theorems, he seemed wary of the fifth and seldom involved it. These and other clues led mathematicians to suspect that the axiom of parallels was not a true axiom at all but a theorem, derivable from the other four axioms. Countless mathematicians down through the centuries tried to derive it. All failed. In retrospect, their failure became a tribute to Euclid's genius in not attempting the feat himself. As the British mathematician Herbert Turnbull wrote, "There is dignity in the way that Euclid left this curious rugged excrescence, like a natural outcrop of rock in the plot of ground that otherwise had been so beautifully smoothed."

By the nineteenth century, enlightened by new developments in symbolic logic, three geometers independently discovered that the reason the axiom of parallels could not be proved was because it was not necessarily so. That slim passway opened onto the noneuclidean geometries.

The German mathematician Karl Friedrich Gauss appears to have been the first to realize that strange new geometries could be constructed by denying Euclid's fifth axiom and substituting another in its place. The results were so weird that even though Gauss was widely regarded as the greatest mathematician of the time, he withheld them for fear of ridicule.

In Hungary, János Bolyai was at work on the problem, against the advice of his mathematician father, who wrote him, "I entreat you, leave the science of parallels alone. . . . I have traveled past all reefs of this infernal Dead Sea and have always come back with a broken mast and torn sail." In 1823 Bolyai discovered that by denying the parallels axiom he could create a new geometry, as consistent as Euclid's but quite different. Nikolai Lobachevski in Russia made the same discovery at about the same time. Bolyai and Lobachevski's geometry was what is now called hyperbolic, meaning analogous to the three-dimensional form of a hyperbola, a shape that resembles a saddle. Bernhard Riemann of Göttingen then composed a spherical noneuclidean geometry. Riemann's Göttingen colleague, Felix Klein, incorporating work by Arthur Cayley of Britain, proved that these new geometries stood on equal footing with Euclid's. Logically, all were valid.

Traditionally science had viewed the natural world, however unruly it appeared, as embedded in a rigid three-dimensional geometry. A flower, an ocean wave or a star could always be mapped upon an unchanging grid of left-right, up-down, and forward-back. Now these new developments raised the odd possibility that three-dimensional geometry itself might be "curved," embedded in a lattice of four dimensions or more. A four-dimensional substructure may or may not be visualizable—Bertrand Russell thought it impossible for humans to imagine, say, a four-dimensional "supercube," while the physicist and physiologist Hermann von Helmholtz felt that children might be taught to do so—but whether or not we can envision such concepts as hyperdimensional geometry or "curved" space, we can think of ways to determine whether nature is built in accordance with them.

Imagine a race of two-dimensional creatures, able to see only frontward, backward and sideways, unable to perceive up and down, who live on the surface of a globe like the Earth. In everyday life, the flat earthlings find their world boundless; they can wander as far as they like in any direction without ever coming to the end of it. But then a flat Magellan undertakes a journey of exploration and circumnavigates the world. Magellan's voyage shocks everyone. The world may be unbounded, they realize, but it cannot be infinite, since an explorer who departed on course due west has reappeared in the east and wound up back where he started.

The flat geometers retire to deliberate the meaning of the flat Magellan's astonishing accomplishment. Ultimately they hit upon an answer. We live, they announce, on the surface of something called a *sphere*. We cannot visualize such an object, because it involves a dimension beyond the two that we can perceive—a *third* dimension—but here, they say, is the geometry of it, quite rational and consistent. This *sphere* is like a circle, but with a dimension added.

To convince flat skeptics, the geometers draw a large triangle in a meadowland, then measure its angles. Bulging subtly with the unseen curve of the globe, the triangle adds up to a total of more than the 180 degrees of a flat triangle. The geometers have demonstrated the presence of an unseen dimension that can be deduced, even if it cannot be perceived.

"There you are," they say. "We live on the surface of a three-dimensional object, a sphere, which to us appears finite but unbounded. Nature is built upon more dimensions than we are accustomed to thinking of."

The creation of the noneuclidean geometries suggested that we three-dimensional humans might be in a situation something like that of the two-dimensional flat scientists. The universe may be built not in three dimensions but in four—or rather, as Mach would put it, nature may respond more readily to inquiry in four dimensions than in three. The idea that the universe might be rooted in a substructure so different from the one we are accustomed to was at first almost frightening, which may help explain why Gauss did not care to see it loosed into the world during his lifetime. But as it happened, the initial impact of the noneuclidean geometries was restricted to mathematicians, philosophers and a few scientists, since no evidence existed to link these abstractions to the real world. Four-dimensional universes might flourish in the pages of books, but Euclid and Newton still ruled Earth and sky. So it seemed until Einstein.

In Zurich and Berlin in the years 1912–16, Einstein worked on general relativity, his theory of gravitation. In addition to the philosophical outlook of Mach's Principle and the power of noneuclidean geometry, he made use of the field concept, which had originated in electromagnetics and which he intended to apply to gravitation. General relativity would do away with the Newtonian concept of gravity as a force acting at a distance, a notion Newton himself had admitted was inexplicable. "Action at a distance," Einstein wrote, "is replaced by the field."

Einstein's vision of the completed theory, at least in outline, may have been clear during these years—his intuitive sense of nature was unsurpassed—but working out the details proved difficult. Noneuclidean geometry is subtle, and Einstein was not much of a mathematician. "Every boy in the streets of our mathematical Göttingen understands more about four-dimensional geometry than Einstein," wrote the geometer David Hilbert, with forgivable exaggeration. "Yet," Hilbert added, "despite that, Einstein did the work and not the mathematicians." Like Gauss, who once cheerfully told a colleague, "I have the result, only I do not yet know how to get *to* it." Einstein knew his goal but was not sure what path would lead him there. Unlike Gauss, he had no confidence that he could find it on his own.

His close friend Marcel Grossman was a devotee of the noneuclidean geometries. "Help me, Marcel, or I'll go crazy!" Einstein wrote. Under Grossman's tutelage, Einstein gained respect for the theories of Hermann Minkowski, whom he remembered from his student days as a cold lecturer who gave him low marks, but now recognized to have been a pioneer in a central idea of general relativity—that noneuclidean geometry could be applied to the physical world by designating time a fourth dimension. When relativity burst upon the world, Minkowski remarked that "for me it came as a tremendous surprise, for in his student days Einstein had been a lazy dog. He never bothered about mathematics at all." Grossman tutored Einstein in the tensor calculus of Ricci and Levi-Civita, the intricate machinery of the quadratic differential equations and the incisive analysis of noneuclidean geometry that had been published sixty years earlier by Bernhard Riemann—whose death of respiratory illness before his fortieth birthday left forever unanswered the question of how close he might have come to relativity himself. "Only the genius of Riemann," Einstein later wrote, "solitary and uncomprehended, had already won its way by the middle of the last century to a new conception of space, in which space was deprived of its rigidity, and in which its power to take part in physical events was recognized as possible."

Newton viewed space and time as separate and absolute. As conceived by Einstein, they are united in a flexible continuum that responds to the presence of matter. The stars and planets wrap the space-time continuum around themselves, so to speak, each sitting in the center of a sort of space-time whirlpool. The commerce we call gravity occurs because objects follow the easiest, most efficient course available to them over the undulations of the continuum. Earth in its orbit glides along inside the sun's spacetime vortex like a roulette ball whirling above the wheel, its velocity balanced against its tendency to slide toward the sun. That tendency to fall "down" the slope of space-time is equivalent to gravity, but no "force" of gravity is postulated. Light beams too follow the dips and hills of the continuum. They trace trajectories that we call "bent," but that is just three-dimensional parochialism talking; they are going just as straight as the shape of space-time permits.

Einstein's accomplishment, of course, was not to propose that space is like a whirlpool and the planets roulette balls, but to create a theory that explained and predicted the behavior of heavenly bodies more accurately and coherently than had its predecessors. General relativity accounted for a peculiarity in the long-term orbital behavior of the planet Mercury that had been inexplicable in Newton's theory, and it has passed every subsequent experimental trial to which it has been put.

The most dramatic of the experimental tests of general relativity came on May 29, 1919, when Arthur Stanley Eddington, Einstein's champion in the English-speaking world, photographed stars near the sun while the sky was darkened by a total solar eclipse. The observing site was a cocoa plantation on Príncipe Island, off the crook of the western equatorial coast of Africa. As the moon cut the sun down to a crescent, rain clouds scudding before stiff winds obscured the spectacle. The sun was still shrouded by clouds when the total eclipse began, plunging the landscape and the astronomers alike into a deep gloom. But then, perhaps due to the drop in temperature, a hole opened in the clouds, permitting Eddington to photograph the stars of the Hyades cluster surrounding the blacked-out sun. He developed the plates in a tent that night, then compared them to plates taken previously of the same star field. If Einstein's theory were correct, stars near the sun should appear to be displaced in the sky by twice as much as predicted by Newton. (In general relativity, half the deflection is caused by the curvature of space, half by the slowing of time in the region of curved space. Were there *no* deflection of starlight, both Newton and Einstein would be proved wrong.) Before Eddington left London for Príncipe, his colleague on the expedition, E. T. Cottingham, asked the Astronomer Royal, Sir Frank Dyson, what would happen should the deflection turn out to be even greater than Einstein predicted. "Then," said Dyson, "Eddington will go mad, and you will have to come home alone." In his book *Space, Time and Gravitation,* Eddington recalled measuring the positions of the stars on the plate: ". . . As the last lines of the calculation were reached, I knew that Einstein's theory had stood the test and the new outlook of scientific thought must prevail. Cottingham did not have to go home alone."

It took months for the expedition to return to England, develop the remaining plates and make certain that general relativity had indeed been verified, as Eddington's preliminary measurements had indicated. The findings were announced in London on November 6, 1919, at a meeting of the Royal Society, beneath a portrait of Newton, the society's president from 1703 to 1727. Four years had passed since Einstein, while putting the finishing touches on the general theory, had found that the theory accounted for the precession of the perihelion of the orbit of Mercury. The experience had so excited him that he could not work for three days. This time, learning of Eddington's observation by way of a telegram from his close friend Hendrik Antoon Lorentz, Einstein could afford to feel more assured. He showed the telegram to a student, Ilse Rosenthal-Schneider, who asked, "What would you have said if there had been no confirmation?"

"I would have had to pity our dear Lord," Einstein replied. "The theory is correct."

Relativity was fresh air to cosmology. So long as space

and time had been regarded as inert and unchanging, merely a stage upon which matter played out the cosmic drama, there had seemed to be no escape from the equally illogical stances of assuming the universe to be either finite or infinite. Relativity demonstrated that in the oceanic reaches between the galaxies, space and time could be distorted until the universe itself might be said to have a shape. Suppose the universe were shaped like a four-dimensional sphere. Such a universe is finite—it contains a finite number of galaxies—but unbounded: an immortal cosmonaut can circumnavigate the universe and never reach a boundary, though eventually he or she will have visited every galaxy. Observers in all the galaxies, looking along the gently curving light beams, see galaxies in every direction; there is no edge in space to which Lucretius' spear-thrower can book passage to perform his experiment.* So the paradox is broken. "This suggestion of a finite but unbounded space," wrote the physicist Max Born, "is one of the greatest ideas about the nature of the world which has ever been conceived."

Many other uniform four-dimensional shapes of the universe seem to be possible. The models fall into three categories—"closed," like the finite but unbounded sphere just described, "open," or hyperbolic, and "flat," meaning euclidean, a middle case between the other two. The real universe presumably can conform to only one of the three categories. The question is, which one?

In 1916, when the general theory of relativity was published, astronomy was unequal to the task of investigating

---

* If the universe was created at a certain time in the finite past, however, we may consider that beginning an "edge" in time. That in fact is what most astronomers now mean when they speak of an edge of the universe. See Chapter 8.

the shape of the cosmos as a whole. Astronomers were just learning to find their way around in our galaxy. They could hardly be expected to test cosmological speculations about the contours of a space-time universe about which they as yet knew nothing more than that it probably was very large. So the cosmologists, for want of observational data, studied not the sky but relativity. There was no lack of satisfaction in this—the beauty of theoretical physics as practiced by an Einstein is comparable to that of a Bach fugue—but something troubled the raptures of the theoretical relativistic cosmologists. The noneuclidean geometries insisted on dancing when relativity called the tune. Relativity, it seemed, implied that the universe expands.

Einstein caught sight of this early but found it incredible. Every astronomer he talked to told him that the cosmos was static and unchanging, that there was plenty of motion *in* the universe, certainly, but no concerted motion to the universe as a whole. Heeding their counsel, Einstein tried to create a relativistic cosmology in which the universe would keep still. The theory resisted his efforts. Einstein's exquisite sense of humor must have permitted him to appreciate the irony of the situation. His essential insight in creating relativity had been to take universal motion seriously. He had rebuilt dynamics to incorporate his realization that, in a cosmos where everything is moving and we lack fixed reference points like those of Newton's "absolute space," any observer's flying frame of reference deserved to rank with any other's. Born of this realization, relativity now seemed to take motion to heart even more fervently than had its author; it saw motion everywhere, put the very cosmos into motion.

For once, Einstein lost confidence and balked. To

force relativity to deliver the static universe that he had been told fit reality, he introduced a new term into the equations that he called the "cosmological constant," designated by the Greek letter lambda. It represented a sort of antigravity operating over long distances. Its effect was to make possible a relativistic model of the universe that would neither expand nor collapse. Einstein published a cosmology incorporating it and soon was sorry. Within two years he had come to regard the cosmological constant as "gravely detrimental to the formal beauty of the theory." Three years later he volunteered that it "constitutes a complication of the theory, which seriously reduces its logical simplicity." By 1931 he had discarded it altogether, branding it the worst blunder of his career.

Meanwhile cosmologists across Europe went on building imaginary universes with relativity. In clusters of chalked symbols, the cosmos bloomed like a flower, warped and curved in an eerie calculus beyond the eye. But it was a cosmos only of the mind. Few expected that reality could be so wild.

Willem de Sitter, director of the Leiden Observatory in the Netherlands, began speculating on new cosmologies almost from the day a copy of relativity theory arrived in the mail from Einstein. An astronomer in the traditional mold, with a white beard and an absentminded manner, De Sitter had a supple and inventive mind. For simplicity, he based his calculations on the assumption that the universe was completely empty. This approach might seem questionable, since we know the real universe contains plenty of matter, but De Sitter meant his model to be an approximation of reality, a mathematical fiction that might have something interesting to say. After all, he would say, the cosmos is mostly space anyway, so what is wrong with a cosmology that is all space? The De Sitter universe was an expanding one so far as its geometry was concerned, but as De Sitter remarked with a grin, it could also be considered static, because there was nothing in it to expand.

Eddington introduced particles of matter into this vacant universe and found that they tended to fly apart. De Sitter at first did not take the implication very seriously. He was preoccupied with the much-debated "De Sitter effect," a mathematical oddity that made distant objects in his model look as if they were receding even if they weren't. This effect was later discounted so far as the real world is concerned, but meantime it caused confusion for a number of researchers, including Hubble.

The first cosmology to suggest clearly that the universe is expanding came in 1922. Its author was Alexander Friedmann, a Russian mathematician who watched the growing clamor over relativity from a position almost on the sidelines, in Petrograd. Like the other cosmologists, Friedmann had no new observational evidence concerning the behavior of the real universe; his technique, like theirs, was to pore over the mathematics of relativity and see whether there was a previously unnoticed lever to be pulled or gear that could be engaged. He was rewarded when he discovered that Einstein, in his "cosmological constant" model, had made an algebraic error, essentially dividing by a quantity that could in some cases be zero. When Friedmann corrected this error, Einstein's hidebound universe glided into motion. Friedmann found that depending upon starting conditions, the universe might expand, or contract, or oscillate, expanding for eons and then collapsing. Defying even Einstein's efforts to restrain it, relativity implied a dynamic cosmos.

Whether the real cosmos acted acordingly was a question only observation could answer. But the observers, generally speaking, knew little cosmology, and many of the cosmologists knew little astronomy. The big new telescopes were in America, while cosmology was still primarily European, its practitioners attached to universities at Leiden, Berlin or Cambridge. Indicative of the gulf separating the two fields was the lack of attention paid Slipher's work at Lowell Observatory on the shift of spectral lines in the light of spiral nebulae. The year Friedmann published the first of his studies of cosmological relativity, Slipher accumulated, though he had not yet published, a list of forty galaxies for which all but four displayed red shifts. This was potent if preliminary evidence that the universe was expanding. The list would have aroused the greatest interest in Einstein, Friedmann, or any of the cosmologists conversant with Friedmann's work, but apparently none of them heard of it. Slipher, it is true, was employed by an unusually isolated observatory, whose general secretary customarily responded to inquiries from outside astronomers by throwing them away. But all American astronomers were, to some degree, detached from the revolution in theoretical cosmology going on across the sea. For want of hard astronomical evidence, Friedmann's expanding universe was dismissed, by those few who bothered to read it, as little more than a mathematical curiosity. Einstein acknowledged it briefly in a polite but unenthusiastic letter. Friedmann died, of pneumonia contracted on a cold-weather excursion in a meteorological balloon, too soon to know the importance of his achievement.

After Friedmann's death his theory sank from view with so few ripples that when a young Belgian researcher approached the same problem only five years later he came across no trace of it and was obliged, unwittingly and laboriously, to duplicate many of Friedmann's calculations, arriving at almost the same results. This was the Abbé Georges Lemaître, professor of relativity and the history of science at Louvain. Like Friedmann, he found in relativity the seeds of universes expanding, contracting or oscillating. But his carefully thought-out cosmologies were not widely read, and, like Friedmann's, they descended into obscurity.

Lemaître anticipated that the universe would be found to be expanding and that this could be confirmed by looking for red shifts in the spectra of galaxies. Probably he had read Slipher's table of red shifts, which Eddington, who had been his teacher, had just published.

The year was 1927, one of the last before cosmologists and astronomers began talking to one another in earnest. A third of the way around the world from Louvain, at Mt. Wilson, Edwin Hubble was sifting from his data a portrait of galaxies rushing from one another. Later, when his discovery linked the two disciplines, he would write a book titled *The Observational Approach to Cosmology,* but at the time he knew almost no cosmology. He appears to have known little or nothing of the work of Friedmann and Lemaître. He had read De Sitter, but only enough so that at first he attributed his red shifts of galaxies to the spurious "De Sitter effect."

With Hubble's discovery of the red-shift-distance relationship, and therefore of the expansion of the universe, "Einstein's theory of general relativity came into its own as a physical theory," the latter-day cosmologist William McCrea wrote. "Hitherto it had predicted minute departures from classical physics. Now it had successfully made the greatest prediction in the history of science." But Hubble did not know the prediction when he published his milestone 1929 paper.

In 1930 news of the expansion of the universe reached the man who had seen it in the sky. Hubble learned of Lemaître's cosmology by reading about it in a semipopular article of Eddington's. Alone in his office Hubble may have danced a jig for all we know today, but he was stonefaced as ever in public. For several years he was reluctant even to admit that his observations evidenced an expanding universe. In 1937 he conceded grudgingly, "Well, perhaps the nebulae are all receding in this peculiar manner. But the notion is rather startling."

3

WHY IS THE SKY DARK AT NIGHT?

*It requires a very unusual mind to
undertake the analysis of the obvious*
—ALFRED NORTH WHITEHEAD

WILHELM OLBERS WAS A German physician who enjoyed sweeping the night sky with a small telescope, looking for comets and asteroids. He complemented this restful way of passing time with a tenacious grasp of orbital mechanics. He calculated the first accurate orbit of a comet, discovering that comets originate in an exurban dump in our solar system out beyond the orbits of the planets. He learned that the asteroids, lumps of rock ranging from the size of a baseball to chunks larger than cities, come from a belt of debris lying between the orbits of Mars and Jupiter. These discoveries made Olbers famous as an astronomer, but he stuck to medicine, exploring the stars as a diversion.

In 1826, at age sixty-eight, Olbers wrote a little paper posing one of the unsettling questions of scientific history. Why, he asked, is the sky dark at night?

In the eighteenth and nineteenth centuries the universe was widely presumed to be infinite. This view deferred the dilemma of how a three-dimensional universe could be finite (if it came to an end, what lay beyond?) and it enjoyed the endorsement of Sir Isaac Newton. If the number of stars was finite, Newton wrote in 1692, their combined gravity would pull them together; they would "fall down into the middle of the whole space, and there compose one great spherical mass." The alternative to that uninspiring state Newton saw to be an infinite universe peppered with infinite stars. "If the matter was evenly disposed throughout an infinite space it could never convene into one mass," he wrote. ". . . Some of it would convene into one mass and some into another, so as to make an infinite number of great masses, scattered at great distances from one another throughout all that infinite space." And indeed Newton's "great masses" sound something like galaxies.

In his deliberations, Newton relied upon euclidean geometry, the one workable geometry known. Twenty centuries of surveying, navigation, building and teaching testified to Euclid's genius, and his system was widely regarded as little less than revealed truth. Euclidean geometry is played on an infinite game board—parallel lines must extend to infinity before they meet, for example—and so, when Newton conceived of the cosmos in euclidean terms, he was inclined to think of it as infinite.

That satisfied almost everyone except Olbers. Why is the sky dark at night?

If there really are an infinite number of stars, we ought to see stars crammed into every niche in the sky. Imagine that a small telescope shows two stars in a given field of view. A more powerful telescope should reveal stars filling the space between them, and a still larger telescope should find stars filling the spaces between *them,* and so on until all we see are stars and no sky whatever. If Newton's model of the cosmos were correct, every line of sight in the sky, no matter how narrow, ought eventually to encounter a star. Prior to Olbers it had been assumed that most of the infinite stars were invisible simply because they were too far away for us to see them. (The apparent brightness of a star diminishes by the square of its distance, so that it one star is a hundred light-years from us and another identical star is two hundred light-years away, the second star will appear to us only one-quarter as bright as the first.) Olbers examined the matter more critically and found that contrary to this assumption, the cumulative brightness of an infinite number of stars should overwhelm the dimming effect of their distance. Olbers concluded that if we do live in a static, homogeneous, infinite universe, as Newton had maintained, the night sky ought to be a blinding sheet of light, bright as the surface of the sun. Yet it is dark.

This is known as Olbers' paradox. Actually much the same point had been made before, in 1744 by a Swiss astronomer, J. P. L. de Cheseaux. Olbers had a copy of Cheseaux's book in his library but apparently never read it, and most scholars credit him with having arrived at the idea of his own. Edmund Halley, known for the bright comet named after him, stumbled on a form of the paradox even earlier, and Johannes Kepler wrote about the paradox in 1610, which was 110 years before Halley. Even Edgar Allan Poe contemplated the paradox of the dark night sky. Shapley mentioned it in 1917, as evidence for his contention that the Milky Way Galaxy might be all there is to the universe: "Either the extent of the star-populated space is finite or 'the heavens would be a blazing glory of light,' " Shapley wrote. ". . . . Since the heavens are not a blazing glory . . . it follows that the defined stellar system is finite."

Olbers considered and rejected two ways to escape the paradox. One was to assert that stars far from the sun get intrinsically dimmer. This only begged the question. Another was to say that the universe is infinite but the number of stars finite. Perhaps the cosmos was composed of the Milky Way and empty space, nothing more. Olbers rejected that possibility because, according to Newton, the Milky Way would collapse if it were true. (Subsequent analysis indicated Newton was wrong on this point, but

since we now know the Milky Way is not the only galaxy—
far from it—the paradox survives.) Olbers finally decided
that interstellar matter probably absorbed the light from
distant stars. This was astute in a sense—Shapley might
have profited by taking interstellar matter more seriously—
but it turned out to be insufficient. Given infinite time,
starlight will heat up interstellar dust and gas until it glows
and the universe will be bright as the sun after all. The
laws of thermodynamics, derived after Olbers' death,
made that clear and renewed the paradox.

Hubble's discovery of the expansion of the universe
seemed at first to offer a solution. Starlight coming from
distant galaxies is weakened by red shifting, darkening the
night sky. But this effect, when calculated, proved insufficient to escape Olbers' paradox.

The resolution of the paradox resides, it now appears,
in the deliverance of time. Stars do not shine forever. They
shine for millions to billions of years, but then they burn
out; looking a long way into space, we may find that our
line of sight encounters not a blazing star but instead a
puff of gas or a dead ball of neutrons that used to be a
star. More importantly in terms of resolving the paradox,
the *universe* has not been here forever, at least not in its
present form. If the galaxies are flying apart from one
another, presumably they once were all crushed together
in a state of tremendously high heat and density, a state in
which stars and galaxies could not have existed. This conception of genesis implies that the oldest stars are not infinitely old but must have been born at a finite time in the
past. If, as contemporary estimates have it, the expansion
of the universe began some eighteen billion years ago,
then, allowing time for the galaxies to form and their stars
to be born and start shining, the first stars should have

begun shining, let's say, fifteen billion years ago. And indeed, the oldest stars known to the astrophysicists are some thirteen to fifteen billion years old.

In a universe where the first stars began shining fifteen billion years ago, the farthest distance a sight line can be extended and still run into a star ought to be some fifteen billion light-years. That's a very long distance, but it's not infinity, and it's not enough to activate Olbers' paradox. It seems that the sky is dark at night because when we look far out into space, we are looking far back in time, to an era before when the stars started shining.

The astronomical term for this kinship of time and space is lookback time.

4

LOOKBACK TIME

. . . all the bells that ever rang
still ringing in the long dying
lightrays. . . .
—WILLIAM FAULKNER

HUBBLE HAD DISCOVERED the expanding universe of galaxies, but his model of the universe, like early explorers' maps of the Americas, was afflicted by contradictions of scale. First, all the other galaxies appeared to be smaller than ours. M33, for example, seemed to be only one-twentieth, and the Andromeda Galaxy perhaps one-fifth, the size of the Milky Way. The picture was one of small galaxies grouped around our galaxy, little dogs ringing a big dog, a situation that would award the Milky Way a suspicious eminence. A second problem was that when Hubble found globular clusters arrayed in a halo around the Andromeda Galaxy, their brightness indicated that either the Andromeda Galaxy was farther away than had been thought, or else its globular clusters were, for some reason, intrinsically fainter than ours. Third and in some ways most disturbing, the Hubble value for the rate of expansion of the universe, if run backward in time, mashed the cosmos into a fireball at an epoch more recent in history than what geologists said was the age of the Earth.

These three discrepancies—the relative dwarfing of galaxies, the dimness of their globular clusters and the unseemly youth of the Hubble cosmos—flawed Hubble's work for years. To his credit, he made no attempt to paper them over. Various theories were proposed to explain the red shifts as caused by something other than velocity, but all were unconvincing, and Hubble, though he flirted briefly with a few of them, refused to be seduced. De Sitter shrugged off the paradox whimsically. "After all," he wrote, "the 'universe' is an hypothesis, like the atom, and must be allowed the freedom to have properties and to do things which would be contradictory and impossible for a finite material structure." Hubble was less blithe. "We face a rather serious dilemma," he wrote. "Some there are who stoutly maintain that the earth may well be older than the expansion of the universe. Others suggest that in those crowded, jostling yesterdays, the rhythm of events was faster than the rhythm of the spacious universe today; evolution then proceeded apace, and, into the faint surviving traces, we now misread the evidence of a great antiquity. Our knowledge is too meagre to estimate the value of such speculations, but they sound like special pleading, like forced solutions of the difficulty."

It took three decades to resolve these contradictions. They arose from several deep-seated errors. Exposing the errors so that the cosmos could be charted more accurately required the building of a larger telescope with which to scrutinize the stars of other galaxies, and the cooperation of astronomers and physicists in a new discipline, astrophysics, that was to crack the secret of how stars work. The central figure in both developments was George Ellery Hale.

When Einstein and Hubble met in 1931 at Mt. Wilson, the walls dividing theoretical physics from astronomy were crumbling. Hale supervised that demolition. He insisted that the three giant observatories he built—Yerkes, Mt. Wilson and Palomar—be equipped with the spectrographs, darkrooms and machine shops needed to make them physics laboratories where stars could be analyzed as well as observed. To publish the results, he founded and edited the *Astro-Physical Journal,* which lost its hyphen as the alliance took. Astronomers in the nineteenth century had been obliged by the constraints of their equipment to content themselves with discerning, as one put it in 1885, *"where* any heavenly body is, and not *what* it is." Twentieth-century astronomy went after the question of *what* the stars and nebulae are. Its style, an aggressive probing of the heavens aimed at projecting physics into the depths of space, is to a considerable degree Hale's creation.

It developed from Hale's love for the sun. As a child he learned that sunlight could be dissected for analysis, a discovery from which he never recovered. He pored over charts of the solar spectrum with the excitement of Columbus scrutinizing Pierre d'Ailly's *Imago Mundi*. At age twenty he invented the first successful spectroheliograph, an instrument that photographs the sun in the light of a single slice of its spectrum. In some wavelengths the sun looked as convoluted as a living cell. In others, it resembled a balloon flecked with ivory speckles. The spectrohelio*scope,* a device Hale built years later, allowed direct and continuous viewing of the sun in many of its exotic costumes.

Hale had a strong sense that we live *in* the universe, that we can learn about the universe by studying things nearby as well as distant. "The sun is a star, the only star whose phenomena can be studied in detail," he told potential contributors to his ceaseless building projects. He had read Darwin and was convinced that physics and astronomy, in cooperation, could trace the evolution of stars as Darwin had that of onions and finches. In 1908 Hale wrote a prophetic sentence, perhaps a century ahead of its time: "We are now in a position to regard the study of evolution as that of a single great problem, beginning with the origin of stars in the nebulae and culminating in those difficult and complex sciences that endeavor to account, not merely for the phenomena of life, but for the laws which control a society composed of human beings."

His career as a telescope builder began one twilight on the veranda of the Powers Hotel in Rochester, New York. Relaxing among colleagues after reading a scientific paper to a gathering inside, he overheard Alvan Clark, a grinder of superlative lenses, remark that the University of Southern California had ordered a 40-inch objective lens from a Paris optical shop and then had been unable to pay for it. The two slabs of cast glass, which if ground to figure and placed together would form the lens of an astronomical refracting telescope, were in storage in Paris. Hale was interested. A 40-inch would be the world's largest telescope, surpassing the 36-inch at Lick Observatory. Considerable money would be needed to buy the blanks, have them ground and build an observatory around them, but Hale, the son of a wealthy father, felt he might be able to raise it. Back in Chicago he made the rounds of potential philanthropists and endured a string of polite refusals. Then he met and won over Charles Yerkes. Years of further haggling and pleading followed, but the recalcitrant Yerkes finally delivered, in bits and pieces, all that he had pledged. Yerkes Observatory was dedicated October 21, 1897, with Hale, age twenty-nine, its director.

The observatory, at Williams Bay, Wisconsin, had hardly begun operations before Hale went to California, camped in an abandoned shack atop Mt. Wilson and tested the night skies with a small telescope. Convinced the site was suitable for astronomy, he gathered backing to build a solar observatory there. The big deep-space telescopes followed, beginning with the 60-inch, its mirror a gift from his father.

Though Hale was physically healthy—he ascended Mt. Wilson at a sort of loping run—he suffered nervous disorders aggravated by overwork, by byzantine negotiations with millionaires and their foundations and by fear that his ambitious projects might fail. When the 100-inch telescope was completed, on November 7, 1917, Hale looked into the eyepiece and saw, instead of a star, a blurred mass of overlapping disks. As it turned out, the mirror was fine. It had been only temporarily distorted that afternoon when a workman let sunlight strike it through the dome slit, expanding its thick glass. But Hale, waiting sleepless in his bunk until 2:30 A.M., when the mirror regained its figure, went through one of many traumas that punctuated his career. By middle age he was beset by insomnia, headaches, numbness in the feet and an inability to read. An imaginary little man appeared to keep him company on sleepless nights. Hale, chatting with this amiable elf, feared that he was going insane.

Under doctors' orders he resigned the directorship of Mt. Wilson in 1922 and attempted to retire. He built a small solar observatory near his home in Pasadena, equipping it with a machine shop and the 12-inch lens he had used as a student. There he invented the spectrohelioscope, outlined the magnetic properties of sunspots and predicted radio storms on Earth set off by eruptions on the sun. He wrote books and articles. He tried to find money to build an observatory in the southern hemisphere. Soon he was back on the psychiatrist's couch.

By then the public's attitude toward science was warming, and Hale's victories began to come more easily. The accomplishments of Hubble, Shapley and others under Hale's direction had aroused American curiosity and pride, while enthusiasm for European scholarship cooled.

Hale himself, who fumed through a year of graduate study at the Physikalisches Institut in Berlin while approving blueprints for the Yerkes Observatory he would return to direct, belonged to perhaps the last generation of American scientists who considered it obligatory to round out their educations in the Old World. Hitler was soon to lend his support to the westward shift of the scientific epicenter. In 1932 Einstein and his wife packed for what she thought would be a brief trip to California. Einstein asked her to take a good look at their villa, because "you will never see it again." Their trip became permanent exile. Nazi storm troopers sacked the house, searching for "weapons" and confiscating a bread knife.

Hale found a receptive audience when he wrote an article proposing a new large telescope. "Starlight is falling on every square mile of the earth's surface," he wrote, "and the best we can do is to gather up and concentrate the rays that strike an area 100 inches in diameter." A large telescope would be expensive, of course, but "if the cost of gathering celestial treasures exceeds that of searching for the buried chests of a Morgan or a Flint, the expectation of rich return is surely greater. . . ." The price of a 200-inch telescope on Palomar Mountain would be six million dollars. With a readiness that surprised Hale, the Rockefeller Foundation pledged the money.

The 16-foot, 8-inch mirror was cast by the Corning Glass Company in New York. In working their way up to it, the glassmakers produced a whole generation of mirrors for new telescopes. Two 61-inch blanks went to Shapley at Harvard, a 76-inch to Toronto, an 82-inch to McDonald Observatory in Texas, a 98-inch to the University of Michigan and a 120-inch, eventually, to Lick. Poured in liquid Pyrex, with a honeycomb back to minimize weight, the

200-inch mirror cooled for a year in a sealed brick igloo. The glassmakers sandbagged the Corning plant against a river flood to stop rising water that would have chilled the igloo and cracked the mirror. Schoolchildren across the country watched the finished blank go by, set on edge in a steel case aboard a custom-built railway car drawn at twenty-five miles per hour. In Pasadena, opticians clad in white smocks ground the blank for two years, were interrupted by the war, then ground for two years more, removing five and a quarter tons of glass to figure the mirror into a parabola that, when coated with a layer of aluminum 1,000 molecules thick, would bounce starlight to a flat, clear focus. The opticians figured the mirror to within two millionths of an inch, an accuracy finer than the vagaries of Earth's atmosphere would ever permit the telescope to realize in practice. It was a matter of pride with them. The dome for the 200-inch and the smaller domes surrounding it were erected in a pasture at 5,500 feet altitude, where tests indicated that the clarity and steadiness of the air, what astronomers call "seeing," were optimal.*

Hale, supervising the 200-inch project, cautioned that he might not live to see it completed. By 1936 his health was failing and he could no longer make the trip up Mt. Wilson. Many astronomers there had never met him, and they regarded him, a biographer wrote, as "almost a myth-

---

* Observatories are built on mountains to raise them above the thick, turbulent lower levels of the atmosphere. In principle, the higher the telescope, the better. In practice, the expense of building mountain roads and the severity of high-altitude weather inspires a willingness to compromise. Mauna Kea Observatory in Hawaii, which may be the highest observatory that will ever be built on Earth, stands at nearly 14,000 feet. There, gale-force winds are common and the air is so thin that higher brain functions are impaired by lack of oxygen. Astronomers quartered halfway down the mountain write themselves childishly simple instructions that they hope their muddled brains will be able to obey when they go up to the dome to observe.

ical being." Invited by Shapley to attend a symposium in his honor, Hale declined in a note that read, "Old and battered fossils may retain a certain antiquarian interest, but in the midst of recent revolutionary advances, they are rapidly outclassed." By early 1938 he could no longer use the small solar observatory he had built for his so-called retirement. He wanted to visit Palomar Mountain and see the big dome under construction but had to keep postponing the excursion for reasons of health. He died on February 21.

Throughout Hale's career the astrophysics he had nourished grew. The story of how stars evolve was being pieced together, in an international effort that many astronomers and physicists consider the superlative achievement of both their sciences.

The early chapters of this story were written just after the turn of the century, when Shapley's Princeton mentor, Henry Norris Russell, and a Danish chemical engineer, Ejnar Hertzsprung, independently discovered a relationship between the colors of stars and their intrinsic brightness, called absolute magnitude. Further research over the years elaborated the Hertzsprung-Russell diagram into a compelling outline of the physical society of stars, in which average stars belong to a gently curving "main sequence," while dwarfs, giants and supergiants occupy branches to either side. The pattern persists when we take into account the color, mass and chemical composition of the many members of the family of stars.

With the rise of nuclear physics came some understanding of what goes on inside stars and why they should fall into the graceful, tree-trunk shape of the Hertzsprung-Russell diagram. The operative mechanism is nuclear fusion, energy released when particles in the nuclei of atoms recombine into configurations of slightly less mass than they started out with. Einstein's famous equation $E = mc^2$—energy equals mass times the speed of light squared—governs the transaction. Because the speed of light is a large number, 186,000 miles per second, even a tiny amount of mass converts into a great deal of energy. Research conducted by Eddington, Robert Atkinson, Fritz G. Houtermans, George Gamow, Carl Friedrich von Weizsacker, C. L. Critchfield, Hans Bethe and others had demonstrated that nuclear fusion is responsible for the furious, enduring heat of the stars and that a variety of fusion mechanisms operate within stars at various stages of their careers, some of them simultaneously.

As pieced together by astronomers and nuclear physicists, the life story of a star goes something like this: A star forms by gravitational contraction of dust and gas. It begins to glow when the temperature and density of its core rise sufficiently for fusion to operate. It enters the main sequence on the Hertzsprung-Russell diagram and settles down there for about 90 percent of its lifetime. That is where the sun is now. Ultimately the star depletes the hydrogen fuel it has been processing at the core. Now it must rely more heavily upon fusion of helium nuclei to survive. Its outer atmosphere expands into a dull red cloud; on the Hertzsprung-Russell diagram it leaves the main sequence and skids into the realm of the giants. When all fusion mechanisms at the core fail for want of fuel, radiation pressure from within can no longer balance the grip of gravity, and the star collapses. The length of time this tale consumes for any given star depends upon how much material went into forming it: A humble yellow star like the sun stays on the main sequence for billions of years before

CARL A. RUDISILL LIBRARY
LENOIR RHYNE COLLEGE

becoming first a modest red giant, then a white dwarf. Larger blue-white stars squander their energy, remain on the main sequence for a much shorter time, balloon into enormous giants and then, theatrical in death as in life, crush themselves through the dwarf phase to become crazily spinning neutron stars or perhaps black holes.

Though the full story of stellar evolution is far from being understood even today, its outlines were sketched sufficiently by the 1930s to be of interest to cosmologists. If pondering the geometry of the cosmos was like mapping the streets of a city, studying about the mechanisms of stars was like learning about the city's inhabitants. A pioneer in applying the new knowledge of the evolution of stars to cosmology was Walter Baade.

Baade was a student in Germany when the Hertzsprung-Russell diagram was first published. He recalled that it "made a tremendous impression on me, because obviously it opened up an entirely new way of studying galactic structure. We could pick out well-defined types of stars . . . and use them to analyze the structure of the galaxy." Eventually he would apply discoveries in stellar evolution to understanding not only the structure of our galaxy but the scale of the universe.

When Baade arrived at Mt. Wilson in 1931, having emigrated from Germany shortly before Einstein did, he guessed, with impeccable intuition, that the problems astronomers were having with their estimates of the age and size of the universe had to do with an insufficient understanding of stars. Stars, after all, specifically variable stars, were the means Shapley and Hubble used to measure cosmic distances.

Earlier, Shapley and Robert Trumpler of Lick had noticed that stars in clusters in our galaxy tended to be idiosyncratic to their home cluster; that is, they resembled each other, in terms of their place on the Hertzsprung-Russell diagram, more closely than they did stars in other clusters elsewhere in the galaxy. This implied, at least in retrospect, that stars within a galaxy are born at differing times as the galaxy evolves and that there might be distinct generations of stars, each with its own peculiarities. Baade chose to investigate the question by examining not our galaxy but another, similar one, a spiral both large enough to resemble our own and nearby enough so that its individual stars could be seen. The obvious candidate was M31, the Andromeda Galaxy. M31 had once been a passion of Hubble's, but he had since moved on to observe deeper space, and Baade, five years Hubble's junior and a newcomer to the observatory, could work on M31 without crowding Hubble.

With the help of Joel Stebbins, a Wisconsin astronomer who frequently observed at Mt. Wilson, Baade first refined to new accuracy the zero-point used in measuring star brightnesses on all photographs. This corrected Hubble's figures slightly upward, though not nearly enough to resolve the dilemma of a universe seemingly younger than the Earth. Baade went on to study the stars of M31, an adventure that eventually consumed twenty years. Resolving stars near the edges of the galaxy's spiral arms was not difficult—Hubble had accomplished this before—but in many ways the most interesting part of a galaxy is its densely populated center. Here, though calculations indicated the 100-inch was up to the task, Baade was unable to photograph individual stars. He tried for years but could get nothing better on his plates than the sort of grainy porridge that had frustrated Hubble in the 60-inch days.

When the United States entered World War II, most of

the Mt. Wilson astronomers were diverted to wartime duty. Hubble was shuttled off to direct a ballistics project in Aberdeen, Maryland, grumbling haughtily that he didn't know what "ballistics" meant. But Baade, born and raised in Germany, had lost his U.S. citizenship papers and found himself classed an enemy alien. Barred from security clearance, he was left alone with the 100-inch telescope. As if in an astronomer's dream come true, he watched from Mt. Wilson as the lights of Los Angeles disappeared into wartime blackout. With the skies darker than he had ever seen them, he again attempted to resolve the nucleus of the Andromeda Galaxy into stars. Again he failed.

As a last resort Baade switched from the customary photographic emulsion, which had a peak sensitivity near the blue end of the spectrum, to red-sensitive plates, which were less vulnerable to contamination from natural background skylight.* The drawback of red-sensitive plates was that they were very "slow," requiring exposures of eight hours or more.

A lot can happen to a large telescope in eight hours. The figure of the mirror changes minutely as its temperature drops during the night. The girders of the tube contract, altering its focal length. The quality of the air varies, moment by moment, blurring star images cumulatively as the exposure goes on. To obtain pinpoint images of stars in M31, Baade charted the weather and scheduled his sessions on nights when seeing was likely to be best, a luxury attainable because he had the telescope almost to himself. On afternoons before observing runs he ordered the dome opened and the slit turned away from the sun, so that the

---

* Human light pollution aside, the sky glows with scattered starlight, faint aurora and zodiacal light, which is sunlight reflected from interplanetary dust.

temperature inside would more nearly equal that outside. When full darkness came he aimed the telescope at the center of the Andromeda Galaxy, opened the plate and guided on a star magnified twenty-eight hundred times. Using a technique of his own invention, he learned to monitor this blob of light for signs of disturbances in the air and quickly change the focus to compensate for them.

"I still remember how confused I was as to what to do," Baade said, recalling the first night he tried his new method. "But you just calm down and wait for a moment of good seeing; and after you have mastered it, it is astonishing how little time you need to see what the situation is, and to give a quick turn to the focus. By training myself in this way I finally learned to handle the method. . . . After shooting was over, it was quite clear that all the precautions had actually been necessary; I had just managed to get under the wire, with nothing to spare." The photographs showed thousands of individual stars in the heart of M31. The margin of resolution was so slim that the *Astrophysical Journal* could not trust the printing process to show that the stars had been resolved, but had to make individual enlargements from the original negatives and bind them into each issue with Baade's paper.

The fact that Baade had been able to resolve the stars in the nucleus of M31 only by resorting to red-sensitive plates indicated that the stars themselves were predominantly red. This impression was reinforced when Baade again tried to resolve the nucleus with blue-sensitive plates, employing the same rigorous techniques he had with the red plates, and again failed. The results fit well with his growing suspicion that the stars in each normal spiral galaxy were broadly divided into two groups. One group, now understood to be older stars, he called Population II; they were red or yellow giants, lacked atoms of elements heavier than hydrogen and helium and were found primarily in the galactic nucleus and in globular clusters. Population I stars, now known to be younger, were yellow or blue and contained a variety of heavy elements that had been seeded into the interstellar medium by explosion of their Population II ancestors. Population I stars are found predominantly in the galaxy's disk; the sun is a Population I star.

What if variable stars also came in these two varieties? When Shapley charted interstellar distances by using Cepheid variables, he had assumed that the two sorts of variables he employed were essentially alike. Some were short-term, oscillating in brightness within a matter of hours. Others varied over periods of weeks or months. Each obeyed the period-luminosity relationship discovered by Henrietta Swan Leavitt—the brighter they were, the longer their period. But only short-term variables were close enough to our sun for Shapley to have measured their distances with any precision. He had extrapolated from these, drawing a curve up through the data points for the long-period variables, and that curve had formed the basis for both his distance calculations and those of Hubble. In 1944 Baade proposed that long-term and short-term variables were actually quite different sorts of stars. He predicted that the long-term variables would prove to be Population II stars, intrinsically much brighter than their short-term, Population I nephews. If brighter, then to look as dim as they did, they must be farther away. If Baade was right, the galaxies were farther apart than in Hubble's figures, the universe had therefore been expanding for a longer time and the time-scale problem would be solved.

Resolution of the question awaited completion of the 200-inch telescope. Shapley predicted that short-term variables in the Andromeda spiral were bright enough to be seen with the new telescope. Baade predicted that even the 200-inch would fail to find short-term variables there, because the Andromeda Galaxy was too far away.

The Palomar telescope with its 15-ton mirror was ready for preliminary tests by December 1947. For first use, in a ceremony known as "letting in the light," it was rigged for visual observation, a rarity for large instruments, where photography and spectroscopy take precedence and observers seldom *look* at anything for long. A little hand spyglass was mounted for use as an eyepiece. John Anderson, who was Palomar project director—he had camped on the mountain on site-selection trips with Hale twenty years before, the two men testing the skies all night while the distant campfires of the Pala Indians flickered below—was now the first to look through the 200-inch.

"What did you see?" he was asked.

"Oh, some stars," Anderson said.

For the benefit of guests the huge telescope was aimed at Saturn, but seeing was bad and the image watery. A globular cluster near the zenith was more of a success: Its stars stood against the black sky like diamond chips in ink.

Hubble waited out weeks of thunderstorms and bad seeing. Then, a few hours before dawn January 31, he rode the small elevator up to the observer's cage to test the telescope. Over the intercom he gave the night assistant the coordinates for Coma Berenices, a good test subject that includes a spray of dim stars in the foreground and a cluster of galaxies in the deep background. Hubble braced himself in the swiveling observer's chair as the telescope heeled over. When the coordinate dials told him he was on target, he focused, checked the field and exposed a plate.

The mirror was still in rough condition, its outer 18 inches left unfinished until opticians could make their final tests, its reflective coating filmed with grime, but the first photographs told Hubble the telescope was going to work. Exposures of five or ten minutes recorded stars the 100-inch could barely reach in a full night. "The tests show quite clearly that the 200-inch opens to exploration a volume of space about eight times greater than that previously accessible for study," Hubble reported. "The region of space that we can now observe is so substantial that it may be a fair sample of the universe as a whole."

When the mirror had been given its final figure, re-coated and installed, Baade got his chance to study the Andromeda Galaxy. If Shapley were right and the galaxy was only 750,000 light-years away, the 200-inch should be able to photograph short-term variables there. But Baade's photographs showed no short-term variables.

Baade withheld his results from publication until he could be sure the telescope was performing to specifications. William Baum, an expert in photocell technology brought to Palomar to gauge the parameters of the new instrument, had confirmed this to Baade's satisfaction by autumn 1952. Now Baade was ready. At the next conference he attended, a meeting of the International Astronomical Union in Rome, he announced that his observations supported his theory that there are two kinds of Cepheid variable stars, that Shapley in measuring cosmic distances by means of the Cepheids had confused the two and that consequently the universe was larger than had been thought. "The error must be such that our previous estimates of extragalactic distances . . . were too small by as much as a factor of two," Baade said. "Many notable implications followed immediately from the corrected distances: the globular clusters in M31 and in our own galaxy now come out to have closely similar luminosities; and our galaxy may now come out to be somewhat smaller than M31." When Baade finished speaking, a South African astronomer, A. D. Thackeray, stood and said that recent observations he had made of the Magellanic Clouds, a prime subject of the southern hemisphere observatories, indicated a dual nature for the Cepheids very much like that put forth by Baade.

Baade's division of stars into two populations was an oversimplification; more than two generations of stars have been born in the long history of our galaxy. But the essence of his point survived and lent credence to George Ellery Hale's vision of cosmic evolution. Astrophysicists today agree with Baade that the old, Population II stars found predominantly in the nucleus and halo of our galaxy and of other galaxies like ours date back to the early days of the universe. They are made of pure hydrogen and helium, the stuff of the early universe. Many more massive stars of their generation spent their nuclear fuel eons ago and exploded, seeding the galaxy's interstellar gas and dust clouds with the atomic nuclei of heavier elements that the old stars had cooked up in their cores. Subsequent generations of stars—our sun among them, and its planets—inherited the heavier elements when they in turn were formed from condensations in interstellar clouds. In the eyes of the astrophysicist versed in stellar evolution, the Earth is alive with cosmic timescapes. A child's helium balloon contains atoms that date from the original material of the universe, but the balloon—and the child—incorporate atoms bequeathed us by Population II stars like those Baade discovered.

Subsequent revisions of the distance scale, primarily by Allan Sandage (who found that Hubble had sometimes mistaken bright gaseous nebulae in galaxies for giant stars) yielded a distance to the Andromeda spiral of 2.2 million light-years. Estimates of other intergalactic distances were increased proportionately. In the reconciled Hubble cosmos, the Milky Way was cast as a large but unextraordinary spiral galaxy, its stars and globular clusters no different from those in other systems. The expansion of the universe appeared to have been going on for at least 10 billion and perhaps as long as 20 billion years, a value in accord with the age astrophysicists assigned the sun, 5 billion years, and geologists the earth, about 4.5 billion years. The universe at last seemed large and old enough for everybody.

For Hubble, relief that the paradox had been resolved was mingled with disappointment that he had not done it himself. "He was upset with the thing," a colleague said, "and he steered away from commenting on it, but he was not surprised. He always called everything he did a reconnaissance."

The expanded distance scale and the advent of the 200-inch telescope prompted astronomers to begin thinking in terms of photographing remote galaxies as they were appreciably long times ago. This was made possible, indeed obligatory, by the fact that the velocity of light is finite.

Imagine light from a distant galaxy traveling a billion light-years and then encountering Earth. It rains down through the atmosphere, a sudden jolt after a billion years of tranquil voyaging, and is absorbed by our pastures and forests, snowscapes, rooftops and seas, a minute addition to the plus side of the planet's energy ledger. One night a fraction of the light is gathered by the curved mirror of a telescope and recorded on a photographic plate. An astronomer develops the plate and looks at it. What does it show? A galaxy, certainly, but not a present-day galaxy. The light, after all, is a billion years old. The astronomer is a cosmic historian, the telescope a time machine, the freshly developed plate a relic ten thousand times older than the pyramids of Egypt. The term for this phenomenon is lookback time: The sky is an image of the past.

When dealing with the large lookback times that pertain between clusters of galaxies, it is not obvious what the "distance" to a remote galaxy "is" within the meaning those words have for us on Earth. Our galaxy and the one we see on a given photographic plate were a certain distance apart when the light set out, but during the eons that passed before the light reached us, the expansion of the universe increased the intergalactic distance considerably. We may with equal justice say that the number of light-years separating us from the remote galaxy is equal to, less than or more than the lookback time. Which answer we choose is as much a matter of philosophy as of science; the concept of distance has only limited usefulness over the cosmological sweep. As a result, the people who deal with the largest distances ever known seldom talk of "distance" at all, preferring to deal with unambiguous observables like a galaxy's red shift, apparent brightness and how large it looks in the sky.

The prospect of photographing galaxies at large lookback times held enormous potential interest to astrono-

mers. It offered a way to study cosmic history directly. If
the universe once expanded much faster than it does today, that fact ought to be ascertainable by studying the red
shift of galaxies so far away that we view them as they
were in an earlier cosmic epoch. If galaxies themselves
looked different in their youth, that too should be observable. As Hubble optimistically forecast, "The combined
efforts of mathematical physicists and observers have reduced the array of possible universes to such a limited
range that it is now possible to predict with confidence
that the type of the actual universe will be identified
within the foreseeable future." To have even a chance to
make that identification was, for young astronomers, an
incomparable prize.

The generation that had led human perception out
among the galaxies was coming to an end. The great opticians who built the California telescopes were dead—
George Ritchey, the irascible perfectionist who called
himself "commanding officer" of Mt. Wilson, but whose
beautiful photographs helped open the world's eyes to the
universe; and Bernhard Schmidt, hard as glass, who
worked without respite for days and nights on end, an
unlit cigar clamped between his teeth, glaring at the mirror on the grinding wheel as if it were a letter bringing bad
news, when instead it was the heart of one of the remarkable Schmidt telescopes that made possible wide-field surveys of deep space. The astronomers were growing old.
Baade accepted a university post in Germany where he
could relax and stay out of drafty observatory domes.
Humason, who remembered waiting on tables at Mt. Wilson, stayed on a few years longer, studying plates
crammed with more galaxies than stars. Shapley retired

from the directorship of Harvard College Observatory in 1952. Hubble died the same year. Immobilized by a heart attack during the final months of his life, he had observed by proxy, dispatching graduate students up the mountain with assignments precise and forceful as battle orders. The brightest of these students, in Hubble's view, was Allan Sandage.

A Midwesterner like Hubble and Shapley, Sandage came from Iowa City, graduated in physics from the University of Illinois and arrived at Caltech in 1948, when the graduate astronomy department there opened its doors. Knowing those doors would lead to the Mt. Wilson and Palomar telescopes, the program's organizers did not open them very far; Sandage and a little band of fellow sufferers flailed through what all remembered as "incredibly hard" courses that seemed to leave no time for sleep. The reward for survival was a chance to complete the work of Shapley, Baade and Hubble and help answer some of the most important questions the twentieth century knew how to ask.

Hubble appreciated the depth of commitment the task would demand and went out of his way to impress his favorite student with a sense that he was passing the torch. Sandage, a self-described "hick from the Midwest," was invited to dinners at Hubble's home with guests of the stature of Aldous Huxley and Igor Stravinsky. "The conversation was intellectual, witty, very Cambridge high table," Sandage recalled. "There was lots of name dropping." If Hubble intended to dazzle, he did. Twenty years later, Sandage still regarded him with awe. "A noble man," he called Hubble. "He interacted with people much as I imagine a god might."

Sandage's devotion to carrying on Hubble's life work

led him to an insularity much like Hubble's. In part this resulted from the envy of colleagues over the publicity Sandage received; he became the man the newspaper reporters called with big questions about how the universe is structured. In part it resulted from his almost religious intensity, his refusal to act like one of the boys, his insistence that what he was doing was every bit as important as it appeared, a quality that earned him the nickname "Super-Hubble." And in part it resulted from specialization; Sandage became very nearly the only person on Earth fully versed in observational cosmology. "Much of what Sandage is doing," said his colleague Jesse Greenstein, "he has been doing for so long that for anybody else just to catch up would take years. And nobody would consider retracing his work anyway, because he is viewed as a man of absolute integrity. I don't know any other field in the world where you can say that of somebody, that he has absolute integrity."

"People have attacked me because I do only one thing," Sandage said in 1975, "but that one thing is to try to figure out how the world is put together. The world is incredible—just the fact that you and I are here, that the atoms of our bodies were once part of stars. They say I'm on some sort of a religious quest, looking for God, but God *is* the way it's put together. God is Newton's and Einstein's laws."

Sandage laughed. "Anyway, I'm a nut, you know. Crazy."

5

THE CREATION OF THE UNIVERSE

Seek simplicity, and distrust it.
　—ALFRED NORTH WHITEHEAD

Sir Arthur Stanley Eddington, who introduced the astronomers and cosmologists to each other, was a Quaker, a mystic, a geometer, philosopher, astrophysicist, popularizer of science and perhaps the greatest astronomer of his age, though it is difficult to say just what age that was. His popular essays on science sound modern today, while his philosophy of science—and he frequently mixed the two, page and paragraph—was widely viewed as a reversion to the eighteenth century.

He was born December 28, 1882, at Kendal, Westmorland. Numbers were his first preoccupation, almost his god. At age four he tried to count the stars in the sky. He mastered the multiplication tables through 24 before he learned to read, and when he *could* read, set out to count the words in the Bible. His widowed mother saved enough of her small income to send him to good schools. In the distilled competition of Cambridge mathematics, he won every prize.

As a scientist he helped diagnose how a star delivers energy from its crucible interior to its surface. He explored the noneuclidean geometries and produced a reformulation of general relativity that delighted Einstein and introduced the theory to the English-speaking public. In his last twenty years he tried to unite quantum physics and relativity with what he called his own "mysticism," his conviction that the universe worth studying is the one within us. "It seems to me that the first step in a broader revelation to man must be the awakening of image-building in connection with the higher faculties of his nature," he wrote, "so that these are no longer blind alleys but open out into a spiritual world—a world partly of illusion, no doubt, but in which he lives no less than in the world, also of illusion, revealed by the senses."

A tall, lean, reticent man, Eddington seems to have been likable, if shy, and a rich conversationalist despite a habit of staring off into the distance; he was literally far-sighted. He stayed at Cambridge most of his life, relaxing by reading detective stories and playing golf. On the golf course he hit a slice and muttered to his partner, W. C. Fields style, "Space seems to be highly curved in this region."

In 1930, speaking to the Royal Astronomical Society, Eddington cited Hubble's recently published paper on the red-shift-distance relationship as evidence that we live in an expanding universe. Eddington and a doctoral candidate, G. C. McVittie, had unearthed Lemaître's forgotten expanding-universe cosmology in the course of a research project, and now Hubble's work had lent it the look of prophecy.* With great self-assurance, Eddington asserted that the universe really was expanding as Lemaître had proposed, and he held to this view even when Hubble, who found it melodramatic, withheld his support. At one point Hubble believed some of his own observations discounted the expanding-universe theory. Eddington coolly dismissed Hubble's worries, and further observations proved Eddington right.

Lemaître, rescued from obscurity and acclaimed as the man who had predicted the expansion of the universe, resumed work on theoretical cosmology.

His immediate problem was to find a plausible expla-

* Friedmann too had predicted that the universe expands, but at this point nobody, including Lemaître, appears to have known of Friedmann's papers.

nation of how the expansion started. In his 1927 model, the one Eddington resurrected, Lemaître had suggested that the universe rested in a static "Einstein state" for an indefinite period before it began expanding. But this static cosmos could be made to seem mathematically reasonable only by invoking the "cosmological constant," and Friedmann had demonstrated that even with the constant included, a relativistic universe might expand. Hubble's results suggested that the constant was unnecessary. Einstein dropped it in 1931, publishing jointly with De Sitter the following year a simple expanding-universe cosmology that did not use it. With the cosmological constant in decline, Lemaître discarded his original, static account of genesis. In searching for a new one, he turned to nuclear physics.

Lemaître proposed that the universe was born as a "primeval atom"—a dense ball of matter as little as 200 million miles in diameter—that disintegrated and flew apart. "Naturally, too much importance must not be attached to this description of the primeval atom," Lemaître wrote. It "will have to be modified, perhaps, when our knowledge of atomic nuclei is more perfect." The important thing was that the universe began violently, in "fireworks," as Lemaître put it.

Eddington, for all his affection for Lemaître, found this new idea distasteful. As an astronomer he was familiar with a universe of steady stars, cold space and galaxies perfect as flowers. He was offended by the idea that everything had once been crammed into a hellish ball. "Since I cannot avoid introducing this question of a beginning," he wrote, "it has seemed to me that the most satisfactory theory would be one which made the beginning *not too unaesthetically abrupt.*" (Eddington's italics)

Eddington believed the laws of nature reside within our minds, are created not by the cosmos at large but by our perceptions of it, so that a visitor from another planet could deduce all our science simply by analyzing how our brains are wired. In Eddington's view, we know physical laws *a priori,* as Kant maintained, although where Kant conceived part of our *a priori* knowledge as inborn, Eddington felt it was derived from experience in observation and reasoning.

Most of us suspect that the world we see is in part genuine and in part distorted, or concocted, by our minds; the question is where the fulcrum stands between internal and external. Eddington put it farther toward the mind than did any other modern scientist. "We have found a strange footprint on the shores of the unknown," he wrote. "We have devised profound theories, one after another, to account for its origin. At last, we have succeeded in reconstructing the creature that made the footprint. And lo! it is our own."

The webs of knowledge link strands unpredictably, as they did for cosmology and astronomy in the 1930s. A scientist may be patient about this process, expecting few final answers within his lifetime. But a mystic, believing the answer is harbored within himself, is less inclined to wait. All he needs, he feels, is to find the key. Eddington looked for truth in the interplay of six apparently irreducible constants of physics—the velocity of light, the constant of gravitation, the charge of the electron, the mass of the electron and of the proton and Planck's constant, a fundamental quantity in modern physics. The key for Eddington was N, possibly the largest number ever conjured up with serious intent by anybody. N stood for the total number of particles in the universe. Eddington estimated it at

10^{79}, or 10 million billion billion billion billion billion billion billion billion. He dropped it into the company of the physical constants, and things started to happen. The radius of the universe in centimeters, R (from Hubble's dwarf figures of the day), divided by the square root of N yielded 10^{-13}, which is close to 10^{-12}, the approximate radius of an electron in centimeters. The square root of N, 10^{39}, was roughly equal to the ratio between the weak (gravitational) force and the powerful (electrical) force that bind atoms together, calculated to be 2.3×10^{39}. Eddington felt that here were clues to the relationship between the very large and the very small, the secret of how, as he put it, an electron knows how large it ought to be.

Another number that fascinated Eddington was 137. Known to spectroscopists as the fine-structure constant, it incorporates the charge of electrons, the speed of light and Planck's constant. Eddington hung his hat on cloakroom peg number 137. The significance of 137 for him resulted from a complicated system, virtually a reconstruction of relativity along new lines, that he evolved in an attempt to bring the cosmos at large into local physics equations. Eddington intended the system to incorporate not only the behavior of atoms but also possible ways they *might* behave (called "degrees of freedom"), as well as the context required to make epistemologically meaningful measurements of them, and even the question of whether they existed (this codified in a mathematical tug-of-war, with nonbeing represented by minus 1). The number of elements Eddington required to describe physical process at a given point in space and time, using this approach, was 136. Well, 136 is pretty close to 137, the fine-structure constant. Eddington attempted to account for the one-digit difference by invoking a term in nuclear physics called the packing fraction, but later investigators who took the trouble to retrace this step found it questionable at best.

Eddington may have been the most lucid writer who was also a cosmologist since Lucretius, but his attempt at unified science and philosophy, published posthumously in a book titled *Fundamental Theory,* was a jumbled landscape viewed by moonlight. It may never be fully understood. Scholarly interest ebbed when both Eddington's N, the number of particles in the universe, and R, its radius, were shifted to higher values by the observations of Baade and Sandage, with the result that many of the coincidences of nuclear numbers with astronomical numbers ceased to coincide. Also, the theory was almost impossible to test. As Poincaré had said, physical theories, "above all, must lead to predictions," and Eddington's did not. Today it is of interest primarily as a historical curiosity.

Yet Eddington was a giant. He shared with Mach and Einstein an understanding that the universe is of a piece. "I only want to make vivid the wide interrelatedness of things," he said. He was set upon the course of his last twenty years in part by contemplating Einstein's equation $E = mc^2$. The world may know this equation as the key to nuclear bombs, but to Eddington its significance lay in the tie it established between the tiny atom and the interstellar world of light rays. The equation "links the universe to the atom," he wrote. He wondered: Why should the nucleus of an atom know anything about the speed of light?

If civilization survives its discovery of $E = mc^2$ for long, physics may well grow toward Eddington's vision rather than away from it. As F. P. Dickson of the University of New South Wales wrote, twenty years after Eddington's death in 1944, "A fair verdict is not that Eddington was 'off the beam,' but that there was as yet no beam and he was trying to show where a beam might be. The attempt was premature but it was beautiful."

George Gamow shared Eddington's conviction that science could learn about the cosmos by studying atoms. But Gamow, an irreverent, broad-brush scientist sufficiently casual about detail that he often neglected to get dates and addresses right, was quite a different man from Eddington. One of his first contacts with Eddington's work came as a practical joke.

As early as the 1920s, less gifted scientists than Eddington had taken to imitating Eddington's cosmic numerology. Well-meaning papers appeared that called attention to previously unnoticed coincidences between this number and that. More practical-minded researchers regarded these efforts as on a par with reading tea leaves or studying the entrails of sacrificed beasts. In 1931, three post-doctoral fellows at the Cavendish Laboratory, Cambridge, wrote an arch paper titled "Concerning the Quantum Theory of the Absolute Zero of Temperature." So carefully worded and closely reasoned was this short note—with references to "degrees of freedom" being "frozen out" in the style of the Eddington disciples—that the editor of the prestigious *Naturwissenschaften* published it without realizing that it was a hoax. A few leading theoretical physicists were taken in too. When the editor learned the truth he was furious. The note concluded, incidentally, that all crystal lattices could be interpreted as constructed according to Eddington's favorite number, 137.

Gamow, twenty-seven years old and an accomplished practical joker, "could not sleep for a week," recalled Max Delbrück, his roommate at the Institute of Theoretical Physics in Copenhagen. "Somebody had outdone him." With coconspirators Wolfgang Pauli and Leon Rosenfeld, Gamow waited for a serious Eddingtonian paper to appear in the *Naturwissenschaften*. Soon one did, an earnest contribution titled "Origin of Cosmic Penetrating Radiation." Gamow and Rosenfeld immediately wrote the journal's editor from separate cities, expressing sympathy that he had again been hoaxed and sorrow that another retraction would now be required. The plot faltered only when Pauli took pity on the man and declined to lend his name to the deception.

Born in Odessa, Gamow left the Soviet Union, where Stalin permitted neither relativity nor Mendelian genetics to be taught in the schools, in 1933. Delbrück described him as "very tall and thin, looking even taller for his erect carriage, blond, a huge skull, and a grating high-pitched voice." Skeptical about religion from an early age, Gamow as a boy had concealed a bit of communion wafer in his cheek, smuggled it home, examined it under a microscope and determined that it was bread and not flesh. "I think this was the experiment which made me a scientist," he said. He liked to tell the story of how the Russian mathematician Aleksei Nikolaevich Krylov had calculated the distance from Earth to the throne of God as nine light-years, on the grounds that when Russian churches directed prayers against the enemy during the Russo-Japanese war in 1905, they had to wait eighteen years (nine years for their prayers, traveling at the speed of light, to reach God, and nine more for His wrath to be transmitted back) before the great earthquake of 1923 struck Japan. Gamow rode around Europe on a motorcycle, loved hiking, sailing, sketching and jokes, loathed boredom. "George had a very acute sense of the value of time," wrote the physicist

William Fowler. "I remember once at a very dull and boring meeting in New York he turned to me and said, 'Sonny Boy, why are we wasting our time here? Let's go have a drink.' Which we did."

Gamow developed a quantum theory of radioactive decay, created a useful model of the atomic nucleus, contributed to astrophysics and to studies that led to decoding the DNA molecule and authored the modern approach to Lemaître's expanding-universe cosmology, for which he is said to have coined the term "Big Bang" theory. His approach to these matters was unfettered by exactitude; he could neither spell nor do arithmetic with any consistency and he had no particular desire to improve, recalling fondly the mathematics professor at Odessa who, when a student pointed out that he had made a numerical error in his calculations, roared back, "It is not the job of mathematicians to do correct arithmetic. . . . It is the job of bank accountants." When Gamow read a paper that interested him he often wrote the author a postcard or a letter, full of minor errors but likely to incise to the heart of the issue. Baade got a Gamow postcard soon after announcing his discovery that stars belonged to two populations. "Please tell me where the lower branch of the color-magnitude diagram joins the main sequence, and I will tell you the age of your Population II stars," it read. Baade replied that not enough about stellar evolution was yet known to make the extrapolation. A second postcard arrived by return mail with a drawing of the diagram. "I have extrapolated the lower branch thus," Gamow wrote. "OK, four to five billion years." Subsequent research traced the curve in the shape Gamow had drawn it, though his figures and Baade's were later revised substantially upward. "That

Gamow hit it so well was an accident," Baade said, "but his remark really contained the whole story of the interpretation of the diagram."

Gamow had studied under Friedmann, the mathematician who suggested that the universe should be expanding, and he was twenty-six years old when Hubble discovered the red-shift-distance relationship. In 1934, while Eddington was calling attention to the connection between Hubble's observations and Lemaître's theory, Gamow arrived at George Washington University, where he was to formulate a new version of Lemaître's cosmology. He had begun thinking of atoms as artifacts of creation.

Most atoms have long lives. Those that form our bodies were in the oceans, soil and atmosphere of Earth before life here began, and they will be here long after we are gone. When the universe was thought to be infinitely old, it was theorized that the interiors of stars, where old atoms are constantly broken down and new ones created, probably acted as the forge where the atoms of our environment were created. Early attempts to take a census of the elements in the universe gave support to this theory. Hydrogen, the lightest of the atoms, was found to be by far the most abundant in the cosmos. Helium, second lightest, is the second most abundant. The amount of energy bound up in an atom's nucleus is in a general way proportional to its mass, and so some physicists concluded that atoms were formed preferentially according to how much energy they harbored. There is more hydrogen around than heavy elements, they argued, because it is easier to make hydrogen. A sort of stellar laziness was thought to rule.

On further study this relationship fell apart. It was found that although the cosmic abundance curve does decline from the lighter toward the heavier elements, it flattens out dramatically less than halfway down the periodic table, at about the position of zinc. Many of the heavier elements exist in approximately equal quantities, even though their masses differ widely. For example, lead has over twice the atomic weight of rubidium, but the cosmos appears to contain about the same amount of lead as rubidium.

Gamow felt that this situation was inconsistent with the theory that the elements were created inside stars. To his mind the cosmic abundance curve suggested the results of a gigantic explosion, like a thermonuclear bomb. Or the Big Bang.

Gamow began to think that if he could reconstruct the conditions under which the elements formed in their observed abundances, he might learn what the universe was like in the first moments after creation. "The relative abundances of various atomic species," he wrote, ". . . must represent the most ancient archeological document pertaining to the history of the universe." The question was how to read the document.

During the war, Gamow was a part-time consultant to the Applied Physics Laboratory at Johns Hopkins. There he met a graduate student, Ralph Alpher, who had just abandoned the subject of his doctoral thesis when a physicist at another school beat him to the result. Casting around for a fresh area of study, Alpher came across a paper that aroused his curiosity. Donald Hughes of the Brookhaven National Laboratory had measured a quantity called the neutron capture cross-section for a variety of atoms and found that this quantity increased sharply throughout the first half of the periodic table of the elements, then flattened out, like an inverted version of the cosmic abundance curve. Neutron capture therefore might have something to do with the way elements were formed in the Big Bang. It might serve as a way to understand the cosmic abundances.

The product of this line of thought was a brash paper attempting to reconstruct events in the fireball of creation. It appeared in a 1948 issue of *Physical Review*—on April Fool's Day, to Gamow's delight. Hans Bethe at Cornell was surprised to find himself listed with Alpher and Gamow as an author of the paper, which he had not worked on; Gamow had added Bethe's name because he thought it would be nice to have a paper on genesis written by people whose names began with Alpha, Beta and Gamma, the first three letters of the Greek alphabet. Bethe had been one of the Cavendish Laboratory students who wrote the Eddington satire seventeen years before, and Gamow may have felt that by including the unwitting Bethe he was finally matching that joke.

The Alpher-Bethe-Gamow theory, like Lemaître's, described the universe as beginning in a highly compressed state, but while Lemaître's "primeval atom" was made of densely packed matter, Gamow's was composed of almost pure energy, with only trace elements of matter. Lemaître's model resembled nuclear fission, the basis for the atomic bomb; Gamow took his from nuclear fusion, the hydrogen bomb. "One should imagine the original state of matter as a very dense over-heated neutron gas," Gamow wrote. He called this stuff Ylem, after the ancient Greek term for the primordial substance from which the world was born. The Ylem cooled rapidly in the moments fol-

Harlow Shapley *(above)* established that the sun lies toward the outskirts of the Milky Way Galaxy. He measured the distances of globular star clusters and found that the clusters are gathered in a giant sphere, centered far from the sun in the direction of the constellation Sagittarius. Shapley surmised that the center of the realm of the globulars was also the center of our galaxy. *Overleaf, left:* The globular cluster M92; each of these ancient associations contains hundreds of thousands to millions of stars, enough to qualify as "a pretty complete 'universe' by itself," Shapley noted. *Following page:* An "open" star cluster; typically younger and less populous than globulars, open clusters are found in the plane of our galaxy rather than out in the sphere of the globulars. *Photos: Harvard College Observatory; Kitt Peak National Observatory; Palomar Observatory*

The Milky Way *(top)* is our view, from within, of the plane of the spiral galaxy to which our sun belongs. Looking outward, we find a strikingly similar plane of dark and bright nebulae demarking the disks of other spiral galaxies, like M104 *(opposite)* and the nearby galaxy NGC 55 *(negative print, above)*. But the hypothesis that we live in one of many galaxies could be verified only when telescopes were built large enough to permit identification of individual stars in other galaxies. *Photos: above, European Southern Observatory; top, Carnegie Institution of Washington, Mt. Wilson and Las Campanas Observatories; opposite, Kitt Peak National Observatory*

Shapley overestimated the size of our galaxy, in part because he underestimated the amount of gas and dust that occupies interstellar space in our galaxy, dimming the light from stars and making them seem farther away than they really are. *Above:* The constellation Orion, embedded in a spiral arm of our galaxy near the sun, is wreathed in the interstellar gas that stretches along the arm. *Opposite:* The Orion Nebula, the middle "star" in Orion's scabbard, a dense region in the Orion cloud. *Photos: above, Carnegie Institution of Washington, Mt. Wilson and Las Campanas Observatories; opposite, Lick Observatory*

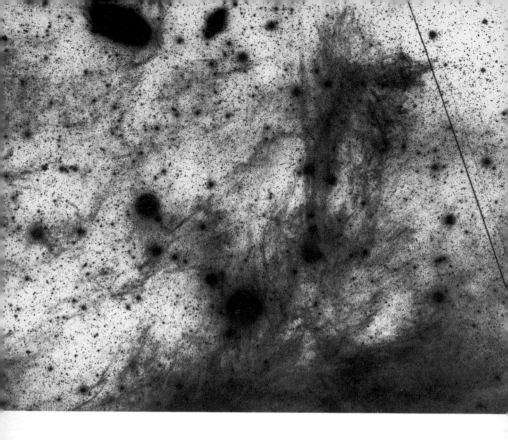

"Galactic cirrus," interstellar clouds that extend out from the plane of our galaxy, can interfere with our view of other galaxies: The large overexposed object at the upper left of the negative print above is the galaxy M81. (For a better view of M81, see eighth page of third photo section.) The straight black lines were traced by the lights of a passing aircraft. *Photo: Allan Sandage, Carnegie Institution of Washington, Mt. Wilson and Las Campanas Observatories*

Where illuminated by nearby stars, dense knots of interstellar gas reveal their full beauty to the eye. *Photos: opposite:* Nebula NGC 2264, *Carnegie Institution of Washington, Mt. Wilson and Las Campanas Observatories; following pages:* Nebula M16, *Lick Observatory;* "Lagoon" Nebula M8, *Kitt Peak National Observatory;* "Rosette" Nebula NGC 2237, *Palomar Observatory;* Rosette detail, *Kitt Peak National Observatory*

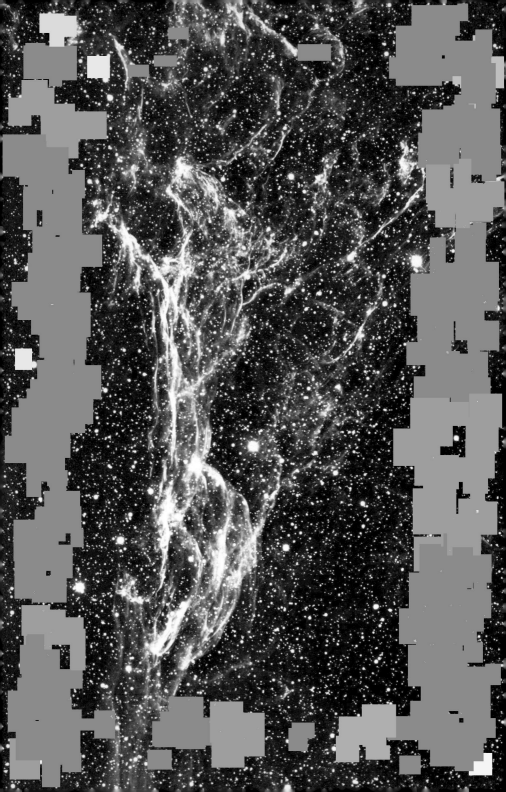

Nebulae like the Helix in Aquarius *(opposite, top, left)* are clouds of gas and dust adrift within our galaxy, while "spiral nebulae" like NGC 4622 *(opposite, top, right)* are galaxies in their own right. Until telescopes were built with which they could be examined in some detail, debate raged over the nature of the spirals. Adrian van Maanen *(opposite, bottom, right)* at Mt. Wilson thought—erroneously—that he had measured rapid circular motions in the spirals, supporting the Chamberlin-Moulton hypothesis *(opposite, bottom, left)* that each was a relatively small cloud of gas condensing to form a star and its planets. Heber Curtis *(above, left)* believed that the spirals were galaxies. As evidence he pointed to supernovae—exploding stars—observed in the spiral nebulae. But initially few believed what we now know to be the case, that supernovae are so powerful they can outshine a hundred billion suns. *Above, right:* A supernova blazes forth in the galaxy M83. *Overleaf:* A giant bubble of debris, ejected from a star that exploded long ago, wafts through our galaxy. *Photos: nebulae, Kitt Peak National Observatory; Van Maanen, Carnegie Institution of Washington, Mt. Wilson and Las Campanas Observatories; Chamberlin-Moulton hypothesis, Yerkes Observatory; Curtis, Lick Observatory; M83 supernova, Milky Way supernova remnant, Palomar Observatory*

lowing the Big Bang; after two seconds it was cool enough for elements to begin forming, a process which trailed off about sixteen hours later. Much later, at a point ten thousand to a million years or so—After the Beginning of Expansion, or "ABE," as Gamow liked to call it—the universe had cooled sufficiently that its energy content dropped below its matter content. It has been dominated by matter ever since.

As a catalyst for provoking thought about creation, the theory was a success. Its spirit is alive in hundreds of early-universe papers being published by cosmologists and physicists today. In terms of its original goal, accounting for cosmic element abundancies, it failed. With the help of two other Johns Hopkins physicists, Robert Herman and James Follin, Jr., and using a borrowed Bureau of Standards computer with an input board that resembled a bowl of spaghetti, Alpher tried to work out the enormously complicated neutron-capture profile required to take the model from Ylem to a universe of elements. Alpher, Herman and Follin found the process blocked at atomic weight 5, and again at 8. The universe contains no stable atoms with those atomic weights; those made artificially in laboratories quickly fall apart. Gamow and his collaborators could find no convincing way for the putative element factory of the Big Bang to have bridged the two gaps. Since the first gap, atomic weight 5, falls just above helium, the researchers were unable, for all their trouble, to cook up any elements other than hydrogen and helium in the Big Bang. Gamow consoled himself that 99 percent of the atoms in the universe are hydrogen or helium, so that he had accounted for the creation of all but 1 percent of everything, but this was cold comfort. The mystery of the cosmic element abundance remained unsolved.

Evidence now suggests that some heavy atoms are indeed made inside normal stars, while other heavier elements, such as gold and uranium, are formed in supernovae and then blasted into space. If so, the gold in our jewelry came from stars that exploded long ago. The idea that heavy elements are cooked up inside stars finds support in the fact that Baade's old Population II stars are poor in metals, as would be expected if they were born before other stars had time to expire and seed the galaxy with heavy atoms.

The cardinal legacy of Gamow's Big Bang theory was a detail scarcely noticed even by its authors. If the universe was once very hot and has been cooling ever since, it ought now to be cold—as it is—but not *absolutely* cold. The energy of the original flash should still be around, though greatly dissipated by the expansion of the universe. In 1948 and 1949 Alpher and Herman, cleaning up Gamow's faulty arithmetic, pointed out in print that the temperature of the cosmos today ought to be about five degrees above absolute zero. The residual energy would take the form of a low-level background radiation coming from all directions at almost equal strength.

It seemed an incidental matter at the time. Few scientists had any idea how to listen for a ubiquitous background radiation; the observation would probably require rockets sent above the atmosphere, or sophisticated radio gear. Gamow, Alpher, Herman and Follin themselves never seriously considered searching for it.

In 1956 Gamow left George Washington University for the University of Colorado, and the team of physicists who had put together the Big Bang theory disbanded.

A decade later, when radio astronomy had developed

to a stature equaling that of optical astronomy, the background radiation was found in one of the most remarkable accidents in the history of science. A certain amount of wistful hindsight ensued. At the Hale Observatories, Allan Sandage recalled once talking briefly with one of the Gamow team—he wasn't sure which one, Follin perhaps—in the early 1950s, who speculated that a rocket might be sent up with instruments to listen for the background radiation. "Nothing came of it," Sandage said, "but for a while there, he had the creation of the universe right in the palm of his hand."

6

THE ECHO OF CREATION

What Is Man?
The Sun's Light when he unfolds it
Depends on the Organ that beholds it.

—WILLIAM BLAKE

THE COOL NIGHT SKY that presents itself to our vision is a guise. Starlight, all light, though rich and beautiful to our gaze, makes up only a sliver of the electromagnetic spectrum through which nature displays itself. The cosmos radiates much of its electromagnetic energy in wavelengths we cannot see. An important excursion into these invisible zones of the spectrum came with investigation of wavelengths much longer than light, those we call radio.

Radio was broadcast artificially well before it was discovered in nature—the young German physicist Heinrich Hertz propagated radio waves in a laboratory in the 1880s, and the Italian Guglielmo Marconi developed a working transmitter at the turn of the century—and as a result the word "radio" retains a connotation of artificiality, like "automobile" or "aspirin." The universe, however, has been in the radio business from the outset and paints its own portrait in those wavelengths as well as in light. If our eyes were tuned to radio (for which purpose they would need be the size of serving platters) the sky would look quite strange—the stars faded except for a few flashing pulsars, the galaxies swollen to several times their optical dimensions, the Milky Way a loom of fire. The universe produces radio radiation in a variety of ways. Once astronomers began to observe at radio wavelengths, they perceived the cosmos as dressed in unfamiliar costumes.

When Hertz discovered radio waves, a few scientists, notably the physicist and spiritualist Sir Oliver Lodge, predicted that there could be such a thing as radio astronomy, but the suggestion ebbed for lack of the technology needed to try it out. The first radio waves to be heard from space were intercepted accidentally by an engineer who was listening for something else.

Karl Jansky, employed by the Bell Telephone Company at their research laboratories in Holmdel, New Jersey, was assigned in 1931 to look into the causes of static on long-distance phone lines. In an open field Jansky built what was in effect a set of exposed telephone cables that could be aimed to determine where static was coming from. The antenna consisted of eight steel rods shaped like squared-off croquet hoops, 12 feet high and mounted in pairs, set on a spindly wooden frame. The assembly, 100 feet long, rested on four spoked airplane wheels and a circular brick foundation. Its receiving cables were wired to an amplifier and moving-paper recorder. Jansky could determine the direction of radio noise by pushing the antenna around in a circle and noting where the recorder hit a peak. A photograph made at the time shows Jansky, in knickers, adjusting the antenna, which looks something like a large unfinished biplane.

Within a few months Jansky had established that noise afflicting long-distance callers fell into three categories. Most came from lightning in nearby thunderstorms, as expected. Additional noise came from distant thunderstorms, and Jansky's finding that this level was higher than anticipated made the experiment a success from Bell's point of view. In addition, Jansky picked up a third kind of radio noise he could not explain, a thin, persistent hiss that did not vary with changes in the weather and seemed to come from one spot in the heavens.

At first Jansky thought the noise came from the sun, something Lodge had predicted in 1894. But as months passed and the sun's position against the background stars shifted, the position of the radio source remained behind, synchronized with the stars, not the sun. By 1933 Jansky estimated that the source was located at roughly the celestial coordinates 17 hours 30 minutes by minus 30 degrees. Consulting a star chart, he found that this coincided with the center of the Milky Way. Somewhat reluctantly, Jansky concluded that Earth was being flooded with radio waves emanating from the heart of our galaxy. It was the first of many lessons to be learned about the strangeness of the radio sky, where distant powerhouses far outshine nearby stars.

Jansky proposed that someone build an antenna intended for astronomy. His antenna worked best at the relatively long wavelength of 14 meters, where resolution, the ability to focus clearly, is inherently low. An antenna built to scan the skies at wavelengths of 1 or 2 meters, he suggested, would resolve cosmic radio sources clearly enough to permit genuine radio mapping of the sky.

Few observatories knew of Jansky's work and none was prepared to finance such a project, but a radio engineer and amateur astronomer named Grote Reber was enthralled by the idea. In the backyard of his home in Wheaton, Illinois, a Chicago suburb not far from the site of George Ellery Hale's old childhood observatory, Reber bolted together the world's first genuine radio telescope. It looked remarkably like its modern descendants. Its 31-foot metal dish reflected radio waves to a receiver suspended at the focus by four outstretched framework arms. A double horseshoe mounting allowed the dish to be pointed in altitude, while the rotation of Earth took care of east-west scanning. The telescope sat behind Reber's house, drawing tourists.

While mapping the radio sky, Reber took a course at

the University of Chicago to refresh his understanding of astronomy. Then he took his maps and charts to Yerkes Observatory and tried to persuade the astronomers there that he had something worth looking at. Otto Struve was the observatory's new director. He recalled that Reber "brought to Yerkes a large stack of tracings, on which his instrument had recorded the intensity of radiation from the sky at a wavelength of slightly less than 2 meters. A definite increase of intensity was shown by a general bulge of the red line, drawn by the recording instrument at the exact time the Milky Way passed over his antenna. The presence of numerous violent and sharp disturbances was an annoying factor, which Reber explained as the result of various electrical appliances, such as a dentist's drill a block or two away, a trolley line of a street in his vicinity or a passing car's faulty ignition."

Struve realized that because radio waves pass through clouds of interstellar dust and gas that block light, the center of our galaxy, forever hidden from optical astronomers, might be observed directly in radio. Most other astronomers, however, were skeptical, even hostile, about this eerie sky that could be observed day and night, where bright stars were invisible, the sun was a dim blotch and the core of the galaxy showed up like the bones of a patient under a fluoroscope. When Reber submitted a paper on his work to the *Astrophysical Journal,* the referees rejected it as unbelievable. Fortunately Struve was editor and printed it anyway, over their objections. Baade, Rudolph Minkowski and Greenstein were among a handful of astronomers whose interest was aroused. "We were beginning to realize," Struve wrote, "that a completely new branch of astronomy was emerging."

The rapid development of radar during World War II,

especially by the British, created a technology radio astronomy would later exploit, and led to a few accidental discoveries as well. A civilian on wartime duty for the British army, J. S. Hey, set up a system for detecting radar jamming by the Germans, was puzzled by a flood of 4-meter to 8-meter-wavelength radiation received in February 1942 and traced it to a solar eruption centered on a large sunspot. That same year a Bell Laboratories researcher discovered thermal, or "quiet sun" radio radiation. A radar array built on the English Channel to detect V-2 rockets picked up the Milky Way background Jansky and Reber had studied. But the scientists involved worked in virtual isolation, their work classified.

Astronomers in the Netherlands, prevented by the Nazi occupation from observing, turned to theoretical work. Jan Oort, who had helped analyze the structure of the Milky Way from photographs and had learned of Jansky's and Reber's discoveries, encouraged one of his students, H. C. van de Hulst, to consider the theoretical side of the matter and see whether he could determine how cosmic radio waves were being generated. Since there is more hydrogen in space than anything else, Van de Hulst concentrated on studying the hydrogen atom. In 1944, just as the occupation was ending, he announced that hydrogen atoms in space theoretically ought to emit radio radiation at a wavelength of 21 centimeters. I. S. Shklovsky in the Soviet Union independently made the same prediction.

Van de Hulst and Shklovsky's work drew upon studies of the magnetic properties of atoms made by nuclear physicists including Bethe, Niels Bohr and Enrico Fermi. At issue was a slight magnetic instability in the hydrogen atom that causes each atom to disgorge a quantum of energy from time to time. This is a very rare event—it befalls a given atom only once in about eleven million years—and interstellar space contains only about one hydrogen atom per cubic centimeter, less than is found in the most perfect vacuum attainable in laboratories on Earth. But there is so much space that the total number of hydrogen atoms, each letting out a single *peep* of energy once every 11 million years, amounts to a continuous radio roar. Calculations indicated the radio radiation should have a wavelength of 21 centimeters. Galaxies are loaded with hydrogen and so presumably could be mapped in detail at the 21-centimeter wavelength. The rate galaxies were rotating could be ascertained, by measuring Doppler shifts of the 21-centimeter line. As the war drew to a close, a race began to detect this important form of radio radiation in the sky.

At Leiden, Oort and a colleague, C. A. Muller, built a 21-centimeter receiver and hooked it up to a 24-foot dish antenna owned by the Dutch Post and Telegraph Service. But their hastily assembled receiver caught fire, and before they could repair it, the discovery was made by two Harvard men, Edward Purcell and E. M. "Doc" Ewen.

Purcell, who was about to win a Nobel Prize for his studies of nuclear magnetic resonance, had worked on radar during the war. He had the expertise required to design a receiver to detect the 21-centimeter signature of hydrogen atoms. When Ewen, a graduate student interested in microwave radio and radar, told him about Van de Hulst's theory, Purcell, as he recalled the incident years later, said, "Hell, we don't need to know that much astronomy, let's see if we can do it."

Their antenna was a wooden horn the size of a bathtub, lined with copper foil and stuck out a window of Harvard's physics laboratory. It cost $400. Passing students threw snowballs into it. Most of the electronic equipment was borrowed. Twin receivers were used, sweeping through the 21-centimeter frequency one after the other, so that spurious radio noise, showing up on both simultaneously, could be ignored, while true hydrogen radiation would reveal itself as a peak when each receiver in turn crossed the 21-centimeter wavelength. On March 25, 1951, Purcell and Ewen recorded 21-centimeter radio. They noted that it reached a maximum when the horn was pointed near the center of the Milky Way.

Six weeks later, Oort and Muller, their charred electronics repaired, duplicated the observation. An Australian team did the same in July.

Radio astronomy developed rapidly thereafter. Large radio telescopes were constructed in Australia, the Soviet Union, the United States and in England, where they were welcomed by British astronomers long hampered by cloudy skies. The discoveries of radio astronomy included exotica like the pulsars—rapidly spinning crushed dead stars—and radio galaxies that blast out too much energy for theorists readily to explain. It was a period when novelty eclipsed retrospection. Few remembered the theories of Gamow, Alpher and Herman in any detail.

We live in an ocean of whispers left over from our eruptive creation, Gamow and his colleagues had said. Nobody was listening.

Gamow was interested in the Big Bang as a way the elements might have been created, but his theory was revived nearly twenty years later by investigators interested not in creating elements but in destroying them. Robert Dicke and his Princeton colleagues P. J. E. Peebles, P. G. Roll and D. T. Wilkinson were looking into the possibility that the universe oscillates, expanding for billions of years, then collapsing into a fireball that might, in turn, erupt as a new cosmos. Dicke was concerned with whether the heat generated in a collapsing universe would be sufficient to destroy all the atomic nuclei, so that each new universe could start afresh without any trace of its predecessor. If so, any question of reading cosmic history prior to the Big Bang would be expelled from scientific concern, since all information from the prior epoch would have been obliterated. If not, studying the traces that survived the Big Bang might tell us something about the universe that preceded it. Dicke and his colleagues estimated that a fireball of one billion degrees absolute—a good "hot bounce," as they called it—would have been required to wipe the slate clean. This avenue led them, from an opposite direction, to the same sort of calculations Gamow had made.

They proceeded with their speculations in 1964 under the pall of what Dicke later admitted was "remarkable ignorance" of Gamow's studies. As they detailed the theoretical conditions of a "hot" Big Bang, it dawned on them, as it had on Alpher and Herman in the 1940s, that energy from the explosion—an echo, so to speak—should pervade the universe today.

Dicke had worked in radio with Purcell during the war, developing the Dicke radiometer, a sophisticated instrument for detecting microwave (that is, very short wavelength) radio radiation. In 1946 he had pointed a radiometer at the sky and determined that any short-wavelength cosmic radiation (which he assumed would come from galaxies) must lie at energies below 20 degrees absolute.* A paper to this effect, written by Dicke and three associates, appeared in the *Physical Review* shortly before Gamow's Big Bang theory was published in the same journal. But the two groups never made contact. Dicke did not have the Big Bang in mind in those days; Gamow, Alpher and Herman did not know that an instrument existed of a sort that could detect its echo.

When Dicke resurrected the idea of a cosmic background radiation in 1964, Roll and Wilkinson set to work building a modern Dicke radiometer to listen for it. The device was small, resembling an old acoustic phonograph, but as elegant as a piece of jewelry, with liquid helium to stabilize the circuitry, and a gold-plated horn pointing skyward. Meanwhile, Peebles began writing a paper describing what they expected to find. The background radiation, if it really originated in the Big Bang, should follow a "black-body" curve—an idealized trace of energy against wavelength predicted by quantum theory. Its temperature would depend upon the total energy of the Big Bang. Peebles was confident that the radiation should still be powerful enough today, some fifteen billion years or more after the fact, to be observable in microwave radio.

That same year, two Soviet astronomers named Doroshkevich and Novikov were reviewing the old papers by Gamow's colleagues Alpher and Herman and came across their background-radiation prediction. They realized the radiation might be observable by radio. They urged that the microwave region was a promising place to look for it and even proposed which of the world's radio antennae

* In the microwave region it is convenient to express signal power as temperature.

was best suited for the job—a horn antenna operated by Bell Laboratories in Holmdel, New Jersey, thirty miles from where, unknown to the Soviets, Dicke and his co-workers were laboring on the same question.

Long radio waves, like big fish, can be caught in loose nets, but microwaves, like little fish, require a tighter mesh, and so the Bell horn was built solidly. It looked like an alpenhorn the size of a boxcar. A scoop at one end admitted the microwaves. The horn funneled them to a focus in a clapboard shed built on stilts. In the portion of the electromagnetic spectrum for which it was built, the Holmdel horn ranked among the most sensitive radio telescopes in existence, but it had been used primarily for satellite communications, not astronomy. The Dicke group had never given it much thought.

Arno Penzias, a young researcher from Columbia, was lured to Bell in 1961 with a promise that if he applied his talents to perfecting the horn as a communication device, he would later be allowed to use it for astronomy. He was joined the following year by Robert Wilson, a Rice graduate who did his graduate thesis on radio astronomy at Caltech.

The two made a good team. Penzias was intellectually aggressive, a generalist interested in music and literature as well as science, conceptually bold if sometimes off-hand about specifics. Wilson was less talkative, more precise, a demon for accuracy; he had the patience to wrestle nominal performance from temperamental, state-of-the-art electronic gear. Penzias was given to enthusiastic monologues, drawing grand interconnections among ideas, while Wilson listened with a smile, interrupting occasionally to correct a figure or define a term.

Bell, with considerable stake in the embryonic communications satellite business, ordered the microwave horn modified so it could talk to the first Telstar satellite. Plans called for Telstar to relay telephone conversations and television signals across the Atlantic by using a Bell antenna in Andover, Maine, on one end and a new antenna in France on the other. But as the Telstar launch date approached, the French antenna was still under construction, and Bell executives began to worry that they might find themselves with an expensive satellite in orbit and only a single facility on the ground ready to communicate with it. "It would be kind of poor to transmit to a satellite and then receive from the satellite in the same place," Penzias said. "You could do the experiment using only the Andover antenna, but there would be something psychologically wrong with it. Somehow, you couldn't bring yourself to gather the newspaper reporters together, send out a pulse, and tell them, 'Well, the signal has been up there and come back.' You want the signal to arrive somewhere else. They weren't sure France would be ready and wanted another receiver prepared just in case."

To do this, the important thing was low noise. The less spurious noise in the receiver, the clearer the satellite pictures would be, and the more promising Bell's debut in the field. To attain the best possible signal-to-noise ratio, Penzias and Wilson fitted the Holmdel horn with a maser (the word is an acronym for Microwave Amplification by Stimulated Emission of Radiation), a device that stored energy and released it when "tickled" by a weak signal of the proper frequency. At the heart of the maser was a ruby crystal embedded with chromium atoms that had to be chilled to nearly absolute zero. Penzias and Wilson jammed four-foot-high yellow steel bottles of liquid helium, temperature four degrees absolute, into the receiver shed and hooked them up by flexible tubes to a jacket surrounding the maser. Valves froze and burst, filling the cramped enclosure with clouds of helium, but when the plumbing behaved the maser worked fine.

As it happened, the French got their antenna into operation by the time Telstar was launched. Signals were beamed across the Atlantic using all three stations, bringing the project to a successful conclusion. Penzias and Wilson now wanted to get on with radio astronomy, using the horn and the new maser. First Bell had one more assignment. Company engineers needed a precise measurement of the horn's sensitivity so that they could be sure Telstar had been performing up to specifications when it relayed signals back to Earth. The engineer's term for the figure needed was "gain." To get the answer, Penzias and Wilson sent a helicopter flying around Holmdel with a small microwave transmitter aboard, aimed the horn at it and recorded the results. Knowing the exact distance of the helicopter and the precise power of its transmitter, they could deduce the gain of the antenna.

The job was mostly drudgery. Penzias and Wilson consoled themselves that if they could establish gain to a high degree of accuracy, they could then go ahead to make some scientifically worthwhile observations of the absolute intensity of astronomical radio sources. This would be helpful to other radio astronomers, who seldom knew the exact strength of cosmic radio objects they observed because they rarely had time to measure the gain of their antennas. So, it was hoped, a little original science might be rescued from an otherwise tedious exercise.

When Penzias and Wilson reduced their data, the results were good enough for the Telstar project but not for astronomy.

The problem was that a low, steady, mysterious noise persisted in the receiver. It showed up on the chart recorders day and night, regardless of where in the sky the antenna was pointed. Such noise had been noticed before. It had posed a problem ever since the maser was installed. It had prompted harsh words between the maser designers, who blamed it on the antenna, and the antenna designers, who blamed it on the maser. Now it had become an embarrassment to Penzias and Wilson. They resolved to identify its source if outside the antenna or clean it up if it were internal; to "exonerate the antenna" or fix it.

They took apart the maser and found it scrupulous. They found two pigeons nested inside the horn. A little of a pigeon's body heat might register in microwave radio. They sent the pigeons by company mail to Bell's offices in Whippany, New Jersey, sixty miles away. The pigeons turned out to be homing pigeons; they were back inside the horn two days later. Penzias and Wilson removed them again. The temperature of the pigeonless horn dropped only half a degree. They took apart the narrow throat of the horn, replaced a few suspicious parts, scrubbed and restored it. This reduced the ceaseless noise only a tenth of a degree. Wilson crawled over every inch inside the horn, masking rivets with aluminum tape. The hiss remained. He and Penzias considered and rejected every natural or man-made explanation they could think of. Months passed. The hiss continued. It came from every quarter of the sky.

"Now here we had a result we knew couldn't be right," Penzias recalled. "You could dispose of any radio source known to exist, if, for no other reason, because all radio sources known at that time radiated more at longer wavelengths than at shorter. Yet here we had this thing in short wavelengths. The only exception is blackbody radiation. And we knew—or thought we knew—there wasn't any blackbody radiation out there, because it was empty space. So we were stuck with a mystery."

So in 1964, three teams of scientists were working in a perfect tripod of ignorance. The Soviets had predicted that if Gamow, Alpher and Herman were right and the universe began in a Big Bang, background radiation following a blackbody curve ought to pervade the sky and be audible to the Bell horn. At Bell, Penzias and Wilson were picking up just such radiation but did not know what it was. At Princeton, a half-hour's drive from Holmdel, Dicke and his team were preparing to search for the noise, without knowing about Gamow, the Soviets or Bell. And Gamow, Alpher and Herman knew nothing of any of their efforts.

Penzias and Wilson prepared to hide news of what they regarded as their failure to account for the noise in the midst of a dry paper relating how they had calibrated the Holmdel antenna. "The early history of radio astronomy was full of incorrect results," Penzias explained. "This is true of any science when it first starts up, but we radio astronomers were terrified of making more mistakes. So we buried it in another paper. Figure the paper went on for twenty pages or so. In the middle of it you mention this residual noise problem. If you're wrong about it, if you overlooked some obvious cause, people won't point to it as the cardinal point in the paper. You're not sticking your neck out so far."

In December 1964, on an airplane returning from an astronomical conference in Montreal, Penzias sat next to Bernard Burke of the Carnegie Institution, in Washington,

D.C., and mentioned his troubles with the residual hiss. Not long afterward, back at Holmdel, he had a call from Burke asking how things were going.

"We still have this damn noise," he recalled having told Burke. "We're pretty sure it really exists, and it seems too important not to publish, but we're sure somebody will shoot us down for putting it out, so we're going to ditch it in this other paper."

Burke said he had just seen a preprint of the paper Peebles had been working on at Princeton, predicting that a background radiation of roughly 10 degrees intensity, left over from the creation of the universe, could be observed with microwave radio telescopes. Penzias might want to read it.

Penzias called Dicke, who sent him a copy of the still-unpublished Peebles paper. Penzias read it and called Dicke again. Dicke drove over and had a look at the horn and the results. He, Penzias and Wilson stared at one another.

To avoid potential conflict, they decided to publish their results jointly. Two notes were rushed to the *Astrophysical Journal Letters.* In the first, Dicke and his associates outlined the importance of cosmic background radiation as substantiation of the Big Bang theory. Peebles' paper had still not appeared, but he was credited for his work. A second note, signed by Penzias and Wilson and titled, "A Measurement of Excess Antenna Temperature at 4080 Megacycles per Second," began:

Measurements of the effective zenith noise temperature of the 20-foot horn-reflector antenna . . . at the Crawford Hill Laboratory, Holmdel, New Jersey, at 4080 Mc/s have yielded a value about 3.5°K higher than expected. This excess temperature is,

within the limits of our observations, isotropic, unpolarized, and
free from seasonal variations (July, 1964–April, 1965). A possible
explanation for the observed excess noise temperature is the one
given by Dicke, Peebles, Roll and Wilkinson (1965) in a companion letter in this issue.

Purcell at Harvard read this modest announcement. "It
just may be," he said later, "the most important thing anybody has ever seen."

Penzias and Wilson found the dimensions of their discovery difficult to absorb. "It never exactly hit us all in one
day," Penzias said. Wilson was surprised that it made the
front page of *The New York Times,* in a bylined story
published May 21, 1965. "Scientists at the Bell Telephone
Laboratories have observed what a group at Princeton
University believes may be remnants of an explosion that
gave birth to the universe," the story began.

"I think reading Walter Sullivan brought it home,"
Wilson said after reading the *Times* account. "It seemed
larger then. More important."

Further observations at other frequencies tended to
confirm that the background radiation followed a blackbody curve—which looks something like a roller-coaster
run, a slow incline followed by a rapid drop—as the Big
Bang theory predicted. The Princeton group got their radiometer operating in six months and observed the noise at
3.2 centimeters wavelength. (The Holmdel antenna was
tuned to 7.35 centimeters.) Its intensity fell on the curve
just where theory predicted. Astronomers at Cambridge
University made a third observation at 20.7 centimeters,
and this too conformed to the expected curve.

These observations were limited to the slow-rising side
of the blackbody curve. The expected peak of the curve

láy at a wavelength that could not be observed from the ground, owing to interference by Earth's atmosphere. Ultimately rockets and satellites were used to chart the background radiation near its peak. But prior to that, several astronomers found an elegant way to make essentially the same observation indirectly, by studying molecules in space. Later, they learned that their work had almost been duplicated in the 1940s, in another of the many coincidences and near-misses that characterized the curious history of how humankind detected the echo of creation.

A variety of molecules has been found floating in clouds in space, among them water, carbon monoxide (which Penzias and Wilson helped detect) and even alcohol and formaldehyde. Just how the molecules got there is not fully understood, but presumably some were formed when two compatible atoms bumped into each other in an interstellar cloud and, like lonely countrymen vacationing where no one else speaks the language, stuck together. Cyanogen, composed of one carbon and one nitrogen atom, was among the first interstellar molecules to be detected. It was observed in the 1930s, its presence betrayed by the faint absorption lines that cyanogen adds to the spectra of stars whose light passes through it before reaching Earth.

Interestingly, many cyanogen molecules in space seemed to be "excited," as physicists say, meaning that their energy levels were boosted above normal. In 1940, W. S. Adams at Mt. Wilson obtained spectroscopic evidence of excited cyanogen molecules in space, and Andrew Mc-Kellar of the Dominion Astrophysical Observatory in Canada then deduced, from the proportion of excited molecules in the cloud, just how much energy was stimulating them. He concluded that the cyanogen was being warmed to a temperature of 2.3 degrees above absolute zero. McKellar and several other astronomers assumed that the energy involved came from starlight. Gamow's theory had not yet appeared, and when it did, no one considered that the interstellar cyanogen might have been heated not by stars but by the residual noise of the Big Bang. Gamow's associate Herman knew of the cyanogen studies but kept them in a separate mental compartment from his cosmological concerns. "I never put the two together," he said. "You know, in retrospect everything becomes clear."

Cyanogen's secret persisted even after Penzias, Dicke and the others published their notes on the background radiation. Penzias later fumed at himself for not having thought of this simple way of watching the echo of creation operating in space. "I *knew* about the cyanogen work," he said. "In fact, when I was working on another interstellar molecule before I turned to the microwave background radiation, I even talked to George Field at Princeton about it—*talked about* the fact that this interstellar cyanogen was at two point something degrees! But I never connected the two at all. Later I felt so stupid."

Field was at Berkeley when he learned that the background radiation had been discovered, and he realized almost immediately that it might be what was heating up the cyanogen clouds in space. At least two other American astronomers, and Shklovsky in the Soviet Union, independently reached the same conclusion, with great benefit for the study of the background radiation. Cyanogen is excitable by radiation at 2.6 millimeters, a wavelength that happens to lie close to the peak of the blackbody curve predicted by the Big Bang theory. By measuring the degree of excitation of interstellar cyanogen, astronomers therefore could, as it were, dip a thermometer into space and sample the radiation. This was soon done. Cyanogen excitation levels were measured in diverse regions of space, and in every case the levels indicated a true cosmic temperature of between 2 and 3 degrees. This was the most powerful confirmation yet of the violent genesis cosmologies of Lemaître and Gamow. Lemaître, seventy-two years old, heard of it in one of his last scientific conversations.

Gamow, Alpher and Herman, the men most vindicated by the discovery of the cosmic background radiation, greeted the news with mixed emotions. They were astonished to find that the discoverers' papers did not credit their foresight. Dicke mentioned their attempt to account for synthesis of the elements in the Big Bang, an effort which had failed, but was silent on the prediction of the background radiation, which now had triumphed. Peebles too was silent. So were Penzias and Wilson. Gamow wrote an indignant letter in his round scrawl to Penzias, listing the occasions when Alpher and Herman had prophesied the background radiation. Gamow had even mentioned it (though at an inflated estimate of 50 degrees, his arithmetic having proven undependable as ever) in his book *The Creation of the Universe,* a best seller that had sold nearly a quarter of a million copies. "You see," Gamow wrote, "the world did not start with almighty Dicke."

Then in his sixties, Gamow occupied a cottage in Boulder, Colorado, wrote popular books on science, answered his voluminous mail with notes, line drawings or poems, drank too much and told friends that each day he felt more strongly that nature was profoundly simple. The dust in space between the stars, he wrote, "has about the same constitution as the dust clouds raised by a bulldozer working on a mountain road." At an astrophysics conference he chaired a session on the microwave background radiation and joked about the way the prediction that his colleagues Alpher and Herman had made had been overlooked: "If I lose a nickel, and someone finds a nickel, I can't prove it's my nickel. Still, I lost a nickel just where they found one." In the last weeks of his life he had recurrent dreams of talking with Newton and Einstein. He died on August 20, 1968.

His younger collaborators, Alpher and Herman, had dropped out of cosmology. Herman went to General Motors, where he applied his skills to examining automobile traffic. He first developed a theory to predict the behavior of cars on single-lane highways, then moved to multilane traffic, then to crossings and complicated patterns like the entrance to the George Washington Bridge. Later, at the University of Texas, he studied the application of physics to the problems that afflict societies. "The future of the world, if it rests on any single issue," he said over lunch in Austin in 1982, "may rest on the ability of science to handle complex problems."

Alpher took a job at General Electric's laboratory in Schenectady, New York. In 1971 he gave up science and moved to an administrative position at GE. "I left science in part because I was kind of disillusioned by the poor scholarship of so many scientists," he said. "Not only Dicke but others. It was discouraging. I would still like to get back into physics at some time in the future, though, if I can still do physics."

7

AN ETERNAL UNIVERSE

According to my opinion, the same force and vigor remain always in the world, and only pass from one part of matter to another, agreeably to the laws of nature, and the beautifully preestablished order. And I hold, that when God works miracles, he does not do it in order to supply the wants of nature, but those of grace.
          —GOTTFRIED WILHELM VON
          LEIBNIZ

In 1948, THE YEAR Gamow published his hot-genesis version of the Big Bang theory, three British astronomers, of whom Fred Hoyle would become the most widely known, created the Steady State theory. This unique cosmology combined skepticism about the Big Bang with an attempt to restore the ancient concept that the universe has forever existed and will exist forever, infinite in time and space. As a theory, the Steady State found little observational support and expired within twenty years. Its importance resides in the fact that it became a rallying point for those skeptical about the Big Bang theory in particular, and about some of the grander claims of science in general.

Thomas Gold, Herman Bondi and Hoyle, all from Cambridge University, were assigned during the war to help develop admiralty radar, at Dunstable, a Canadian Air Force base northwest of London. Bondi and Gold rented a cottage near the aerodrome. Rent was low because the cottage sat next to a bomb dump. Canadian attack bombers would return from dawn raids along the French coast carrying live time-fused bombs they had been unable to drop due to flak damage or failure of the release mechanism. The bombs were nondefusable. Anxious crews carted the ticking bombs to the dump and left them to explode. When they did they blew out windows in the cottage. The dull thud of the bombs going off alerted the local glazier, who selected glass panes of the correct size, rode out to the cottage with them on his bicycle and replaced the broken windows. He billed the Canadian Air Force on a monthly basis. In this house two or three nights a week, Bondi, Gold and Hoyle met and discussed the nature of the universe.

To the extent that the three could be assigned roles, Bondi, a brilliant mathematician, handled the analytical end of their talks, while Gold, naturally skeptical, questioned assumptions. Hoyle was the instigator. Tough-minded and unpolished, Hoyle had been raised in a textile valley in the North of England and climbed to Cambridge on a ladder of scholarships. He retained a distrust of what he felt were "establishment" scientific ideas, among them the Big Bang theory. "Hoyle was the driving force, always asking questions of us," Gold recalled. "How could Hubble's observations be understood? He was full of Hubble in those days." Unsatisfied with the answers Lemaître and Gamow favored, and prompted by the time-scale problem Hubble had encountered, Hoyle groped for some new way of interpreting the expansion of the universe.

After the war the three kept up their talks at Cambridge, usually in Bondi's rooms at Trinity College. One evening Gold fell into a short monologue and heard himself asking, as he later recalled it, "Do we really know the universe must have come from a beginning? After all, we had to create the matter then, at the beginning. Why not create it a little at a time?" Suppose matter were being created continually in space, Gold said. Then the universe could expand yet never thin out. New galaxies, formed out of new matter, would fill the voids between the old.

Gold later thought that the idea might have been planted in his mind at the local movie theater when he saw *Dead of Night*, a film that begins and ends with the same scene, lending it a circular quality. Whatever generated the idea, Gold felt he had hit on something. Hoyle and Bondi were, at first, less impressed. Bondi offered to explore its mathematical ramifications, expecting to find contradictions. Within three weeks he reported that instead, it worked rather nicely. Continuous creation, it was obvious from the outset, violated the law of conservation of energy, a foundation stone of physical science that states that the total amount of matter and energy in any closed system (including a closed universe) never changes. But so, it could be argued, did any of the fireball cosmologies that talked of "creation" in an instant long ago. And Bondi found that the extent to which continuous creation theory violated the law was not so great as it had first appeared. While new atoms theoretically appeared out of nothing, old galaxies meanwhile receded to near the edge of the observable universe, their light so red-shifted it could hardly be detected. So the matter-energy content of the *observable* universe might remain constant even though new matter was being created out of nothing.

Hoyle proposed that they work up a continuous-creation, Steady State cosmology and publish it jointly. But as the three talked they found themselves divided over whether to concentrate on the continuous-creation aspect of the theory, as Hoyle preferred, or upon the Steady State aspect, favored by Bondi and Gold. The result was that Hoyle wrote one paper, Bondi and Gold another. Hoyle finished and submitted his first, to the irritation of Gold, who had been talking about the theory as his own idea. But the journals Hoyle first sent his paper to rejected it, so Bondi and Gold won the little derby. Their published paper was dated July 13, 1948, Hoyle's August 3.

The Bondi-Gold paper was philosophical in tone, light on mathematics and strongly influenced by the work of Edward Milne, an Oxford mathematician who had turned to cosmology in the early 1930s. Always a mathematician at heart, Milne esteemed deduction. The inductive method prevalent in science, with its emphasis upon observation and experiment, struck him as primitive. He hoped science would get over it. "The more advanced a branch of science, the more it relies on inference and the fewer the independent appeals to experience it contains," he wrote. Milne developed a theory of kinematic (that is, based upon motion) relativity, intending to replace Einstein's formulation with one deduced entirely from the passage of time. The theory was not well received, but elements of it colored cosmological discussion for years. Particularly important to Bondi and Gold was what Milne called the "cosmological principle."

The cosmological principle held that models of the universe must allow everyone in the universe the same general view. There may be no privileged observers. Cosmology was not to repeat the pre-Copernican mistake of placing humans in the center of things or otherwise attempting to account for appearances by special pleading. The principle did not say that the universe must look *exactly* the same from all perspectives; localities naturally vary. Observers near the core of spiral galaxies, if there are any observers there, bask in skies blazing with stars, while others, circling stars expelled from their parent systems, journey in the blackness of intergalactic space; large spirals like ours afford their residents a bounteous Milky Way, while others, the small, irregular galaxies, are paltry as spilled salt. Milne's point was, rather, that the large-scale look of things from every point in the cosmos must in general resemble ours, that in any plausible model of the cosmos our perspective must be assumed ordinary. As such, the cosmological principle remains more or less a rule of the game.

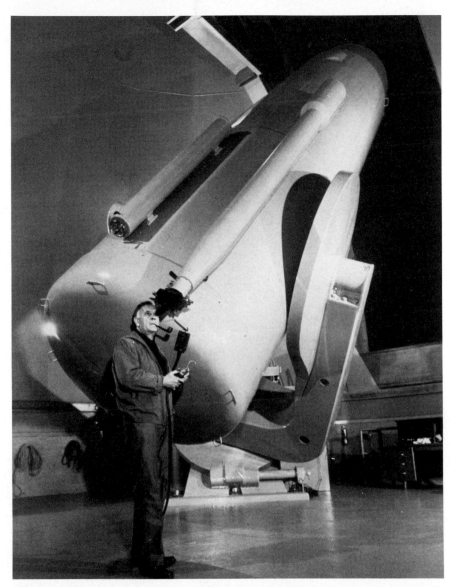

Edwin Hubble, discoverer of galaxies and of the expansion of the universe, poses at the 48-inch Schmidt telescope atop Palomar Mountain. This publicity photograph amused astronomers who knew Hubble as an indifferent observer who was far better at planning observations and interpreting the results than at actually using a telescope. *Overleaf:* Hubble with his cat, Nikolus Copernicus, and *(following page)* the photographic plate on which, in October 1923, he found a Cepheid variable star in the Andromeda Galaxy. Since Cepheids in our own galaxy were known to be very bright, the apparent dimness of this Cepheid meant that the Andromeda spiral must lie far beyond our galaxy. *Photos: above, California Institute of Technology; overleaves, Carnegie Institution of Washington, Mt. Wilson and Las Campanas Observatories*

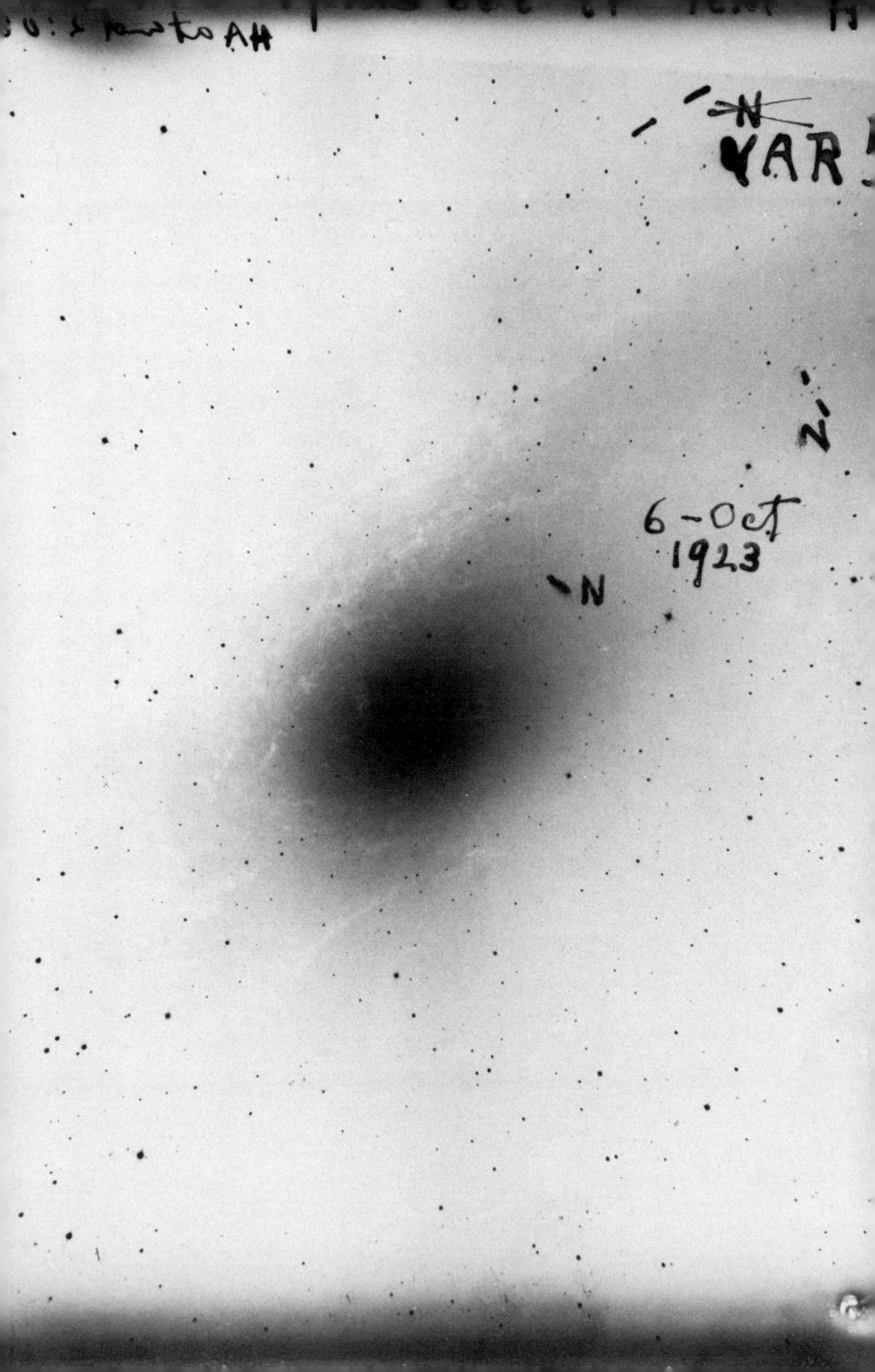
N
VAR!
N
N
6-Oct
1923

The Andromeda Galaxy, sister spiral to the Milky Way, was a focus of Hubble's search for recognizable stars in other galaxies. Stars in the Andromeda spiral could not be positively identified in photographs taken with smaller telescopes *(opposite)*. But the 100-inch telescope made possible photographs *(above)* that indisputably resolved Andromeda into an "island universe" of stars. The stars scattered across the field of view lie in the extreme foreground, within our galaxy. *Photos: Palomar Observatory; Carnegie Institution of Washington, Mt. Wilson and Las Campanas Observatories*

Plate Number   Remarks

313   Shapley reported near diffuse variable nebulae here. Nothing; the sorts how Shapley object is probably an accident.

316   SBc ??

317   fa variable.   PMT 7:55 - 8:55

318   3 exposures. Centered ... south HA at ... 8 ... SA 136 end at 0:55 W

319   2 double exposures ... south ... 0:18 W ... end at 0:44 W   SA 136 end at 40 W.

320   Scale plate as power of (√2) - 362 - 256 - 181 - 128 - 905 - - 64 - 45 - 32 - 225 - 16 - - 115 - 8 - 56

321   2 small faint neb. Sc & Sb.   1 asteroid near edge   - - 2 · 15 · 1.

322   Sc tilted

323   No changes in form & variability   1 PM *   1 var

324

325   Cassegrain focus.

326   8:35 - 9:05 P.M.

327

328

329   8:55 - 9:55 P.M.T.

330   Using Searss visual filter & 350 plate.

331   Nova suspected

332   SBb

333   R Co ... faint

334   7:30 - 8:30 P.M.T.

335   Confirm nova suspected on H 321.4 ... on the date (H221.H), three stars were found 2 of which were novae and 1 proved as variable. Later identified as a Cepheid - the first recognized in M31

336   R near minimum seen rather brighter than last minimum

337

338   Seems to be a giant planetary. Star in center very blue @ 10.5 vis mag. Could focus with ...

339

340

Above: A page from Hubble's observing notebook, noting his epochal discovery of a Cepheid variable star in the Andromeda spiral. Opposite: Milton Humason, who started at Mt. Wilson as a mule packer and worked his way up to staff astronomer. He collaborated with Hubble, providing observational skills Hubble lacked. Overleaf: Negative print of a photograph taken by Humason of the galaxy M81, ten million light-years distant. Photos: Carnegie Institution of Washington, Mt. Wilson and Las Campanas Observatories

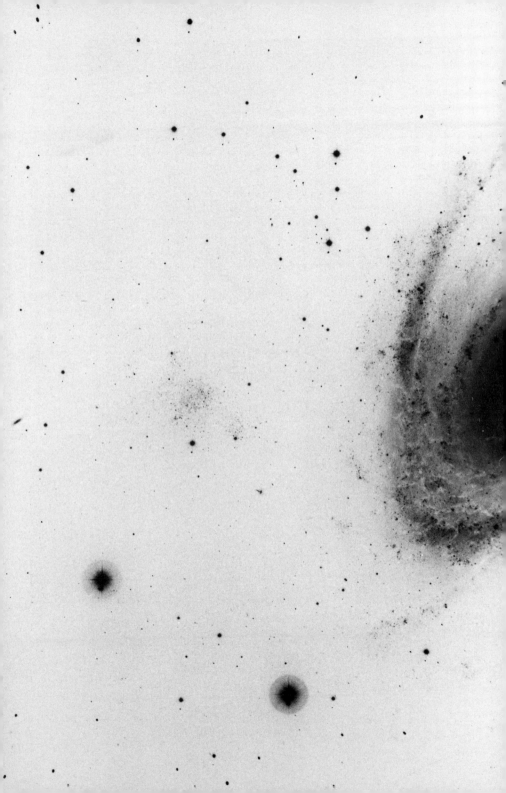

Allan Sandage, Hubble's pupil, carried on Hubble's studies of the structure of galaxies and the expansion of the universe, as well as conducting research in stellar evolution and observational cosmology. *Above:* Sandage at the 100-inch telescope at Las Campanas Observatory in the Chilean Andes. *Opposite:* Galaxy M83, photographed by Sandage from Las Campanas. *Preceding page:* Satellite galaxy of M81, photographed by Sandage with the 200-inch telescope at Palomar Mountain. Negative prints like these are favored by astronomers because they reveal greater detail. *Photos: Sandage, Douglas Kirkland; M83, M81 satellite, Carnegie Institution of Washington, Mt. Wilson and Las Campanas Observatories*

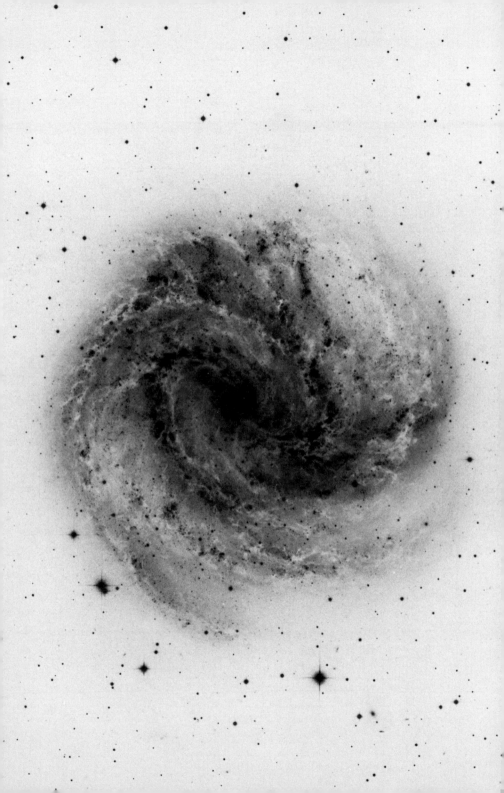

Above: Dedication ceremony on Palomar Mountain in 1948 for the Hale telescope, its 200-inch-diameter mirror the world's largest precision collector of starlight. *Opposite:* The galaxy M101, approximately 24 million light-years distant, photographed by Allan Sandage using the 200-inch telescope. *Following pages:* Galaxy Centaurus A, classified "violent," "peculiar," home to a thousand billion suns, photographed with the Hale telescope; Virgo A, with its over *three* thousand billion stars and its halo of perhaps a million globular clusters; interacting galaxies M51 and NGC 5194, photographed with the 4-meter Mayall telescope at Kitt Peak National Observatory in Arizona. *Photos: Hale dedication, California Institute of Technology; M101, Carnegie Institution of Washington, Mt. Wilson and Las Campanas Observatories; Centaurus A, Palomar Observatory; Virgo A, M51, Kitt Peak National Observatory*

Bondi and Gold took Milne's principle a step farther. Since relativity requires that we regard the galaxies as inhabiting a continuum of time as well as space, they asked, why confine the cosmological principle to space alone? Why not demand that the universe present the same aspect to observers not just in all places, but also at all *times*? Their answer was to postulate what they called the "perfect cosmological principle." Our overall outlook on the cosmos, they maintained, ought to be shared by observers in all quarters of the universe and at all times in its history. Evolving-universe theories violate the perfect cosmological principle because they imagine the universe was once very different, before the galaxies formed, and they imply that in the future, whether the universe expands forever or eventually collapses, things will be much different than at present. Only the Steady State theory, where new galaxies constantly form to fill unfolding space between the old, satisfies the principle its authors devised.

Hoyle preferred a less immodest approach, concentrating on the continuous-creation facet of the theory. In his original paper, as well as in later versions he developed with J. V. Narlikar and in a revision by William McCrea, Hoyle's theory added a "creation" term, designated by the letter C, to Einstein's general relativity equations. In such a theoretical universe, a feedback mechanism balanced the rate of expansion and that of creation of new matter. In part because his theory contained at least a bow in the direction of relativity and therefore incorporated a theory of gravitation, as Bondi and Gold's did not, Hoyle became the leading advocate of the Steady State theory among scientists. His gift for popularization—he wrote books on astronomy and science-fiction tales—gained him the same role in the public eye.

As a wholly new cosmology, the Steady State had its points. Lacking a fireball genesis of the sort that Gamow thought might have created the elements, it looked to stars as the forges of the heavier elements, and this hypothesis turned out almost certainly to be correct. Indeed, it was skepticism about the Big Bang that led Hoyle to one of his most lasting contributions to astrophysics, a landmark 1957 paper with William Fowler and Margaret and Geoffrey Burbidge detailing how chemical elements may be created inside stars.

There was, of course, no evidence that so much as a single particle had appeared out of empty space, but this, curiously enough, was not a telling argument against the theory. The universe is mostly space, and the rate at which creation would have to proceed in order to satisfy the theory was slow—only a couple of atoms need appear per year in an area the size of a cubic city block. A physicist could scrutinize an empty liter jar for a million years, and if not one atom appeared in it, he would still not have made a case against the Steady State theory.

As a philosophical question, which side to take was almost a matter of taste. If evolving universe cosmologists regarded the idea of atoms popping into being out of empty space as implausible, the Steady State advocates found the Big Bang equally so. The Big Bang, after all, requires us to accept that the cosmos was once crushed together at great heat and density, and we are permitted to inquire neither how large this universal furnace was, because no caliper could exist to measure it, nor how hot it was (Gamow's calculations deal with the first moments *after* expansion began), nor what it was doing before expansion commenced. Milne called the fireball a "grand irrationality." Hoyle admonished that "it is against the spirit of scientific inquiry to regard observable effects as arising from 'causes unknown to science,' and this in principle is what creation-in-the-past implies." And Gold cautioned that continuous creation ought not to be rejected merely because it seems to defy common sense: "We often think we can make intuitive judgements on matters which have quite demonstrably never entered into our experience. There is hence no virtue in timidity in the choice of hypotheses. They may look drastic and offend our dwarfish preconceptions, so long as they do not offend any evidence. . . ."

Gold's remarks suggest something of the scientific climate in which the Steady State theory arose—a climate of growing skepticism about the Big Bang theory. The infirmities suffered by this, the reigning cosmology, troubling when it was in its infancy, were becoming increasingly unsightly as the theory aged. In 1948, astronomers had known of the expansion of the universe for almost twenty years, but, owing to the undersized distances Hubble and others had assigned to the galaxies, the universe as modeled by the theory was unbelievably young. One had to be a serene thinker indeed not to be bothered by a cosmology that made out the universe to be younger than the Earth. Internal contradictions in a prevailing view invite radical alternatives, and the Steady State theory was radical.

The climate changed in the early 1950s, when Baade, then Sandage, discovered the sources of substantial errors in Hubble's figures. The corrected distance scale suggested a far older and roomier cosmos and thereby did away with the contradictions that had fertilized the growing popularity of the Steady State. Thereafter, the Steady State had to rest on its own merits and not upon the deficiencies of its chief rival. Like any good theory, it was open to test by observation. These tests it failed.

Hoyle's predicted value for the deceleration parameter—the rate at which the expansion of the universe is slowing—was not borne out by observation. A Steady State universe should contain many very old galaxies, some infinitely old; but as the 200-inch telescope was used to sample the cosmos to new depths, no extremely old galaxies were found. When Penzias and Wilson discovered the cosmic background radiation, fulfilling an almost forgotten prediction of the Big Bang theory, the Steady State received a mortal wound. By 1970 it had few remaining advocates.

Enthusiasm for the Steady State, however, translated into enduring skepticism about Big Bang theory, which was attacked as a simplistic model that put blinders on its adherents. The trouble with the Big Bang theory, it was said, was that it encouraged astronomers to view data in its terms, to accept what fit and explain away what did not. "The guy who goes to the telescope at night and knows in advance exactly what he's going to observe, that guy drives me right up the wall," said Halton Arp, one of the dissidents. "Mavericks disturb things, rock the boat, cause trouble, and some, a majority, are dead wrong. But the ones who are right are very, very valuable. They know it is only from observation that you are going to learn anything new." So the Steady State-Big Bang debate evolved into a fresh version of an ancient question: To what degree does theory bias observation? On the one hand, some theoretical foundation must be admitted if a scientist is to make any observations at all; as Einstein remarked, "Science without epistemology is—in so far as it is thinkable at all—primitive and muddled." On the other hand, excessive adherence to theory can be crippling; Ptolemaic cosmology, which put Earth at the center of the universe, eventually hardened into the dogma that Galileo was persecuted for questioning. The schism among cosmologists was as much a matter of philosophy as of science, and it persisted when the Steady State theory faded away. In the 1960s it manifested itself as a debate over quasars.

In 1959, thirty-three astronomers were polled about cosmology. "Did the universe start with a 'Big Bang' several thousand million years ago?" they were asked. A third said yes. "Is matter continuously created in space?" Fifty-four percent said no. "Is a Gallup Poll of this kind helpful to scientific progress?" All thirty-three answered no.

8

THE RED LIMIT

"O dark dark . . ."

—T. S. Eliot

ASTRONOMY HAD PENETRATED to nearly the edge of the universe—or did so in the opinion of many astronomers—by the early 1970s. Newspaper accounts of the development raised as many questions as they answered—how can the universe have an edge?—but what had happened was perhaps less exotic than the phrase suggested. The Big Bang theory (or theories) predicted that there should be a distance in space-time beyond which no galaxies or stars, no cosmic lanterns of any sort, appear. For years this prediction attracted little attention among observers, because the distance to the hypothetical edge was far greater than that to which telescopes could photograph galaxies. Then in 1960 quasars were discovered. Red shifts in the light of the quasars indicated that these objects were extremely far away and extremely bright—bright enough that existing telescopes could see them across billions of light-years. By the early 1970s quite a few high-red-shift quasars had been found, but none with red shifts greater than a certain value. That value corresponded to the time that the Big Bang theory implied galaxies first formed. It appeared that more distant quasars were not being seen because there were none. The quasar cutoff point approached "the edge of the universe."

It is an edge not in space but in time. The Big Bang theory envisions that the explosion that started the universe expanding was followed by an interval of time that elapsed before galaxies and stars began to form. Vast clouds of dust and gas congealed into galaxies, which in turn subdivided into stars. The first light to grace the cos-

169

mos may have come from stars, from the furiously burning cores of young protogalaxies or from some other source since extinct; the important point is that prior to the advent of this light, except for the heralding flash of the fireball, the universe abided in darkness. How much time darkness prevailed has proved difficult to estimate; theorists guessed it might have been as little as 10 million or as much as several billion years. In any case, astronomers today, probing back into time with their telescopes, ought to find cosmic history exhaustible. If the first stars, quasars, or whatever began shining, say, fifteen billion years ago, then fifteen billion light-years is the limit to which astronomers will be able to see anything. When they look farther they will encounter only darkness; they will be looking back to a time before there *was* anything to see.

The edge should show up at about the same distance in all directions, for all observers anywhere in the cosmos. Nobody lives "at" the edge, or any nearer to it than anyone else, for the edge belongs to history. It is distant from us in the way that the living Socrates is distant from us, as part of the irrevocable past. We can see deep into history because the universe is so large: Had Socrates, on the day of his trial in Athens, stepped aboard a spaceship and sped away at nearly the speed of light, we could look through telescopes today and see him, alive, on board that ship. His image would be a ghost, in the sense that it would be a long beam of light that might or might not be said to have a living man on the other end, but that is just another way of saying Socrates belonged to the past. His face, in the telescope, would be with us *now*.

So it is with galaxies. We see a young galaxy or quasar on the "edge" of the universe, speeding away at nearly the velocity of light, as it was when its light first began to shine. In the billions of years since, it presumably has evolved into a middle-aged galaxy like our own, but we cannot yet see the result of that evolution: It will take billions of years longer for light to arrive with the news. The citizens of that evolved galaxy, turning their telescopes our way today, see the quasar or galaxy or protogalaxy or whatever our galaxy was, billions of years ago, when the universe was young.

The four-dimensional universe of space-time is difficult to imagine, in that it is difficult, if not impossible, to visualize. If we try to fix a four-dimensional sphere or hyperbola in our mind's eye, it eludes us. When we try to envision the universe of space-time, at one moment it seems we are in the "center" of the cosmos, at the next that we are out near the "edge." To our perceptions either is correct, because there is no three-dimensional "center" to the universe, any more than there is a two-dimensional "center" on the surface of a sphere for the flatlanders. The question is not just "where" we are in the cosmos, but *when* we are. The Big Bang theory's answer to this question is that we live roughly eighteen billion years after expansion began, something more than fifteen billion years since the lights came on.

As the universe expands, observers in any galaxy find that the other galaxies are moving away at speeds proportional to their distances—the more remote each is, the faster it recedes—and this recession can be measured by determining the Doppler shift of light from the galaxy toward the red. (It is equally correct to imagine that the galaxy we observe is stationary and *we* are the ones moving so fast, since there are no fixed points of reference.) Relativity theory requires that the speed limit be that of light: Remote galaxies may be seen receding at velocities very close to that of light but never equal to it. If the universe were infinite in time and space, the farther we looked the more galaxies we would see, and we could press on indefinitely and would always find more galaxies, with velocities ever closer to, but never achieving, that of light. In a Big Bang universe with a finite history, however, we would at some distance encounter the "edge," and beyond it see nothing. In all versions of the theory, the edge is far away in space and time—lookback times on the order of fifteen billion light-years are cited—and so the edge will be found where galaxies are receding at something over 90 percent of the speed of light. Observing the putative edge of the universe therefore is a matter of seeing objects at great distances with large red shifts.

Existing telescopes cannot record galaxies at that remove. For years the distance record belonged to Rudolph Minkowski, who, in 1960, in a virtuoso performance on his last 200-inch observing run before retirement, obtained spectra for a cluster of galaxies receding at nearly half the speed of light. According to prevailing estimates, this would put their distance at some seven billion light-years in terms of lookback time. Electronic image-amplifying equipment later improved telescope performance, but the range to which galaxies could be observed remained less than half the estimated radius of the universe.*

That the universe should, early in its history, have thoughtfully provided cosmic beacons brighter than galax-

---

* By "radius" I mean the time since expansion began, expressed in light-years—e.g., eighteen billion light-years if the universe is eighteen billion years old. The true radius of the universe will differ somewhat from this simple definition according to the degree of curvature of space-time, which is in turn a function of the cosmic mass density.

ies for the astronomers of today to observe seems an unexpected favor. But that is just what quasars appear to be. Their red shifts indicate that virtually all quasars are remote. If so, they belong to an earlier cosmic epoch. Their tremendous luminosity suggests that they may represent a violent phase galaxies go through when they form. Whatever the cause, the result appears to be that we can make out quasars all the way to the edge of the universe.

The discovery of quasars was made possible by radio astronomy. In the late 1950s, as new radio telescopes came into operation and techniques for using them improved, radio astronomers were able to map more accurately the coordinates of radio sources in the sky. Optical astronomers were so interested in having a look at the radio sources that a lively traffic developed in "bootleg" radio coordinates, passed along informally before being published. If an optical astronomer received a new set of coordinates from a friend at a radio observatory and aimed his telescope at the indicated spot in the sky, he stood a good chance of finding something unusual, to his delight but to the frustration of colleagues not favored with the information. In the competitive world of astronomy, where only a few dozen observers gain regular access to the large telescopes, friendships were cemented or eroded according to who was or was not favored with fresh radio plots.

In 1960, Tom Matthews, a radio astronomer at Caltech, was working on refining the coordinates of several radio sources observed by British astronomers at Cambridge and at Jodrell Bank, a radio telescope installation in Cheshire, England. These particular sources intrigued Matthews. Each appeared concentrated, limited to a very small patch of sky, which suggested that they might be far away. Matthews made up a list of ten such sources. Later all would prove to be quasars.*

Matthews managed to get an improved radio fix on one of the sources, 3C48 (for number 48 in the Third Cambridge Catalogue of cosmic radio objects). He passed the coordinates along to Allan Sandage, who used his next session on the 200-inch telescope to photograph the sky in the indicated spot. When Sandage developed the plate he found that the location of 3C48 coincided with a blue point of light of medium brightness, far too dim to be seen with the unaided eye but brighter than many of the things he was accustomed to photographing at Palomar. "I took a spectrum that night and it was the weirdest spectrum I'd ever seen," Sandage recalled. "I took the spectrograph off the telescope and put the photometer on, to check its colors, and the colors were different than any object I'd ever seen. The thing was exceedingly weird."

Several other compact radio sources were located with sufficient accuracy that optical telescopes could find them, and the list of known quasars grew to perhaps half a dozen. Each was analyzed with spectroscopes and the other tools of optical astronomy. None seemed to make sense. Several of the quasars showed clear spectral lines, but the lines fell in places along the spectrum that failed to coincide with the characteristic "fingerprint" of any known element. Equally difficult to explain was the fact

---

* The name, coined by a NASA physicist, stands for Quasi-Stellar Radio Source. It is doubly misleading—quasars are "quasi-stellar" only in the sense that they were at first mistaken for stars, and most are not radio sources—but by the time quasars were better understood it was too late to change it. Professional enthusiasm for the name never ran high, and when the Second Texas Symposium on Relativistic Astrophysics affirmed its use in December 1964, the vote was twenty yeas and forty abstentions.

that though quasars looked like stars, they displayed bright emission lines in their spectra instead of the dark absorption lines characteristic of stars. Emission lines had been recorded in galaxies that radiated powerfully in radio, but they were not usually found in stars. To add to the puzzle, Sandage, continuing his observations of 3C48, learned that its light varied in brightness over a matter of weeks.

The realization that the light of quasars was extremely red-shifted eluded astronomers for two years. Most felt, with varying degrees of assurance, that if quasars looked like stars they probably were stars, even though previously no stars had been known to radiate powerfully in radio. "I tried very hard to identify the spectral lines of 3C48," Sandage said, "but I was baffled. I had been raised to believe, with Hubble, that the only things that could have large red shifts were galaxies. So my mind was completely blocked [to the possibility of substantial red shifts in the light of quasars]. Both Tom [Matthews] and I assumed the quasars were stars."

A case study in this mental block was that of Jesse Greenstein, a transplanted New Yorker who had headed Caltech's graduate astronomy department since it opened in 1948.

Greenstein was something of a patriarch to the Mt. Wilson and Palomar astronomers; nearly half the observers winning time on the major California telescopes in the 1960s had trained at Caltech under his guidance. His favorite time of day was lunch in the Athenaeum, the Institute's wood-paneled, cubiform dining room, when he could sit and talk with his ex-students. "They're all my boys," he said in his rasping growl. "To me it's one of the finest, most exciting things in the world to sit at lunch with everybody and see my foster children being so bright. It's a wonderful place."

His specialty was stars. "I'm a person who understands the spectra of atoms at various stages of temperature and pressure. A lot of things can be diagnosed about stars and galaxies from their atomic spectra and I'm something of an expert on it. I've studied, in particular and in great detail, how you get the composition of stars from their spectra. I can look at the spectrum of a star and tell you its age, its composition, its temperature and pressure. Any puzzle in spectroscopy I think of as a challenge." When Sandage, after trying for months to decipher the spectrum of 3C48 without success, mentioned the matter to Greenstein, Greenstein followed the bent of his expertise and came up with a way to explain the quasar as a star.

"I kept talking to myself, and finally convinced myself it was some kind of strange star, a thousand light-years or so away, the result of a supernova," he recalled. "I was able to find a physical condition in which atoms of helium and oxygen could be stripped of their outer electrons and produce spectra of ionized helium and five times ionized oxygen. You could understand these lines if you said these things resulted from a supernova."

Once, over lunch, Greenstein and a fellow Caltech astronomer, Maarten Schmidt, came achingly close to the correct solution, when they talked about whether the lines could be those of familiar elements like hydrogen and helium red-shifted to unfamiliar positions. But they dismissed the suggestion as, in Schmidt's words, "wild speculation." Later, when red shifts were found to be the answer, Greenstein was amused to reflect that he had treated quasars almost as a psychological projective test, stamping them with the imprint of his own preoccupation with stars. "It was a classic story," he said, "of inhibition of one's creativity by knowing too much."

It was Schmidt who finally got at the truth about the quasar spectra.

Then one of the youngest of the Caltech astronomers, Schmidt had been born in Groningen, Holland, in 1929. He dated his interest in the stars from a summer night when he was twelve years old and looked through a small telescope at his uncle's home in Bussum. He soon made a telescope of his own out of a cardboard toilet paper roll and a magnifying glass he found in a junk pile. "It works, to my tremendous surprise," he wrote his uncle. He began observing each clear night with the telescope poked out of a third-story window. At Caltech, where he joined the staff in 1959, Schmidt was recognized as a skilled observer, someone who could extract top performance from a telescope. Most astronomers do their own observing but few enjoy it; as Greenstein said: "When you're in the observer's cage of the 200-inch, the telescope turning and the stars going by, it's romantic, beautiful, marvelous, but on the other hand it's uncomfortable, tedious, bone-cramping and you get sleepy." Schmidt, if he did not exactly love telescopes, at least felt at home with them. "Something about the sight of the 200-inch is deeply impressive," he said. "There's just something about it. Observing with any large telescope is essentially an adventure, even though your feelings usually are not exhilaration but of worry that you've forgotten something or are doing something wrong."

Schmidt began observing quasars in May 1962, and was able to obtain the first spectrum of 3C286. It revealed only a single line at a seemingly irrational location. He wrote, for the *Astrophysical Journal Letters,* what he privately called "a complaining note, saying, 'Here's this thing and we can't understand it.'" Another quasar, 3C147, turned up a few lines but they too defied Schmidt's attempts to explain them. 3C196 came out blank.

Then Sandage, acting on another of Matthews' radio coordinates, photographed the quasar 3C273 and discovered a thin jet, or spike, protruding from it. A similar jet had been found jutting from the nucleus of M87, a galaxy in the Virgo cluster that Sandage had observed many times himself. M87 was one of the brightest known galaxies and was also a strong radio source. It was definitely a galaxy, and so 3C273, which resembled it somewhat, might be a galaxy too. But first impressions die hard. Schmidt and Sandage did not imagine that the starlike quasars belonged in the distant habitats of the galaxies.

Schmidt obtained a spectrum of 3C273 and found a rich variety of lines that made no sense to him. He had just turned thirty-two.

On February 5, 1963, Schmidt sat in his Caltech office, a room with high ceilings, a gray wool carpet, tables buried under books and papers, and a large, uncluttered oak desk. Sunlight filled the tall windows. He held the 3C273 spectrum, a sliver of film smaller than a stamp, pinched between the thumb and forefinger of his left hand. With his right hand he idly sketched the spectral lines on a yellow notepad. The British journal *Nature* had requested a note on the new quasar, and Schmidt was preparing what he anticipated would be another inconclusive, "complaining" report. He felt vaguely irritable. He was coming down with a cold. He set down the pen, picked up a magnifier and peered again at the frail spectrum on the film.

As Greenstein recalled it, he was walking down the hall past Schmidt's open door when he heard Schmidt call him back. Schmidt told Greenstein that he thought there was a red shift of 16 percent in the light from 3C273. "Impossible!" Greenstein said. Quasars all look like stars, he had invested considerable effort explaining them as stars, and stars do not have enormous red shifts. But Schmidt showed Greenstein the pattern he had suddenly perceived, how four of the spectral lines could be interpreted as a well-known series of lines emitted by hydrogen atoms and displaced by 16 percent toward the red end of the spectrum. Intervening lines complicated the picture, but Schmidt could already see how most of them could be explained.

It was one of those moments when scientific investigation overlaps into exploration. A 16 percent red shift, if produced by the expansion of the universe, would put the quasar at a distance of over a billion light-years.

The two men covered paper and blackboard with equations, talking excitedly in the shorthand vocabulary of astrophysics and spectroscopy. Then a curious thing happened. Greenstein felt as if a vault door deep within his subconscious mind had swung open, revealing that it held the red-shift solution. John Bolton, a radio astronomer, had suggested to Tom Matthews that another quasar, 3C48, might have a 37 percent red shift; Matthews had mentioned the idea to Greenstein. Greenstein had discounted it, but now he found that his subconscious had retained not only the hypothesis, but even the exact figure. Greenstein told Schmidt that the red shift in the light of 3C48 must be 37 percent. They retrieved the 3C48 spectrum from Greenstein's files and examined the displacement of its lines. It was 37 percent. This second

quasar, apparently, was four billion light-years distant—
very nearly the most distant object theretofore recognized
by human beings.

There was a brief celebration, then Schmidt returned
to his desk and spent the rest of the afternoon reassuring
himself that he had overlooked nothing, was not somehow
making a mistake. By nightfall, he and Greenstein, satisfied that quasars did display large red shifts, had devised a
preliminary proof that nothing else could account for the
positions of all the spectral lines. The only way known for
quasars to show such spectra was for them to lie at great
distances in an expanding universe.

Schmidt drove home and told his wife, in his Dutch-
accented English, "Something horrible happened at the
office today." He saw her shocked expression and realized
he had chosen the wrong word. "I mean, something wonderful," he said.

"Your hope, when you start a career in science, is that
you will discover something," Schmidt said ten years later.
"Of course, there are many kinds of discovery. You may
find a new comet in the sky, in which case your name goes
on the comet and it's well established that you made a
discovery, but that is actually an incidental thing, almost
like being a witness to an accident. This quasar business
was very different. It totally changed our insight and outlook upon the future study of the universe. Even now that
we've been with quasars for years, we haven't recovered
from the surprise. I simply can't believe it."

Whatever they were, quasars seemed to be some ten
times brighter than galaxies. To search out quasars would
be to probe the universe to unprecedented depth. A staff
astronomer at the Hale Observatories, Schmidt expected

to be allotted twenty to twenty-five nights per year on the 200-inch. This time was his most valued resource. He invested most of it in hunting quasars. For the next three years—riding the telescope while it paced the stars and the music of Bach and Telemann played on the sound system in the dome—Schmidt accumulated specimens of quasars. By 1964 he was seeing farther in space and time than anyone had before. In 1965 he recorded quasars receding at up to 80 percent the speed of light.

Other astronomers joined in the search, among them E. Margaret Burbidge, a British astronomer who was one of the few women to take a place in the first rank of observers.* She and E. Joseph Wampler, inventor of an electrical image-amplification device she irreverently called the Wamplertron, measured the spectrum of a quasar receding at more than 90 percent the speed of light. At Kitt Peak Observatory in Arizona, C. Roger Lynds and Derek

---

* Mrs. Burbidge, who was responsible for converting her husband, Geoffrey, from physics to astronomy, found when she arrived at Mt. Wilson that she could use the telescopes only by adhering to the fiction that she was his assistant. George Ellery Hale, annoyed by clashes between astronomers' wives at Yerkes Observatory before the turn of the century, had decreed that Mt. Wilson was for men only, and in the 1950s the tradition endured, as did others, including one by which the astronomer scheduled for the 100-inch telescope sat at the head of the table at mealtimes. Barred from Mt. Wilson's dormitory (known as the Monastery), the Burbidges were obliged to stay in the same cabin Hale had found atop the mountain when he first climbed it in 1903. When Margaret Burbidge protested, she was told the real reason for discouraging her was that the observatory had only a single bathroom. "Who cares about that in a dome?" she asked. "I mean, if it's only a matter of a *loo*. . . ." Ultimately she was admitted to equal status by the California observatories. When the Queen of England, in naming her director of the Royal Greenwich Observatory, severed the title from that of Astronomer Royal for the first time in its 300-year history, Burbidge resigned in little over a year and returned to the University of California at La Jolla. In 1971 she declined the American Astronomical Society's Annie Jump Cannon prize for outstanding women astronomers, on the grounds that "because of the small number of women in the field," it would "not be surprising if we all in our turn are selected."

Wills identified a quasar red-shift equivalent to 89 percent the speed of light. In 1973, Mirjana Gearhart of Ohio State and R. F. Carswell and P. A. Strittmatter at Steward Observatory in Arizona found a quasar, OH471, with a velocity of 90 percent the speed of light. Quasars with even higher red shifts were discovered in the 1980s. Light from these objects may be said to have been traveling, depending upon the model of the universe one chooses, for some twelve to fifteen billion years.

The span of this odyssey did not especially impress those who achieved it. Accustomed to big numbers, the astronomers were more interested in the spectra themselves, in seeing unusual patterns of lines from above the high-frequency end of the spectrum shifted down into the visible region. Margaret Burbidge found it "terrifically exciting to think that we were actually seeing that far," not into the universe, but "into the ultraviolet—into regions of the spectrum normally blocked by the earth's atmosphere, that we thought we'd never see until space telescopes were developed." To Schmidt, this new spectroscopy was "for me much more exciting even than the discovery of the red shift of 3C273, though this has been appreciated by practically nobody. All those new lines had to be identified. It was a bootstrap operation. Nobody could tell me what lines to look for; the quasars had to tell it."

Talk of the edge of the universe arose not because quasars were so far away, but because none was found farther away. By the time that red shifts had been measured for about two hundred quasars, only a few were so distant as to be receding at 90 percent the speed of light, and almost none was farther. If there were many quasars more remote, any of the major telescopes ought to have been able to photograph them. But as time passed and the list of observed quasars grew, the sharp cutoff at about 90 percent held, and a number of astronomers began to feel it probably was genuine, not an accident or an observational effect. "If the cutoff is indeed real," said the Hale Observatories annual report for 1971-72, "we may be looking back in time so far . . . we have seen 'the edge of the world' in space and time." *The New York Times* on April 8, 1973, ran the headline, "Men Report Seeing Edge of Universe."

The expansion of the universe began, according to Sandage's measurements of galaxy distances and red shifts, some eighteen billion years ago. Stellar evolutionists set the ages of the oldest observed stars in our own and nearby galaxies at fifteen or sixteen billion years. The level of the cosmic background radiation suggested a cosmos fifteen to twenty billion years old. The quasar cutoff came at sixteen to seventeen billion light-years. Either the universe was born in fire eighteen billion years ago, or it was performing an adept impersonation of one that was.

But what are quasars?

Or rather, what *were* they? Few if any remain in the modern cosmos. Schmidt and Sandage analyzed the accumulated data and found a pattern of quasar evolution with cosmic time. Locally, and therefore recently, no quasars were observed. Within a few billion light-years only a few were known. At distances such that their light had been traveling roughly half the history of the universe, eight to ten billion light-years, quasars were more common by a factor of one hundred. At still longer lookback times their frequency rose three hundred times, then one thousand times. Then at a point equal to about 90 percent of the age of the cosmos came the cutoff, and suddenly there was none. "Apparently quasars were numerous in the early universe," Schmidt said. "They evolved, died and now are very rare. Probably they evolved into something else. I think they probably are the nuclei of galaxies, but mind you, that's speculation."

It remained for future researchers to decipher the phylogeny of quasars, galaxies and an array of mysterious denizens of deep space bearing names like Seyfert galaxies, N-galaxies, Markarian galaxies and BL Lacertae objects. Very likely, all might eventually be understood as stages in the evolution of galaxies. Some quasars and BL Lacertae objects, for instance, when observed carefully, prove to be surrounded by a halo that displays the spectrum characteristic of a spiral galaxy—persuasive evidence that they are, as had been suspected, the violent nuclei of galaxies. Seyfert and N-galaxies, for their part, exhibit the characteristics of quiescent quasars. But while connections have been found linking quasars and galaxies, it remains to be seen whether the incredible violence of a quasar phase is something that has happened to most galaxies, or is a sort of disease caught by relatively few. Only about 1 or 2 percent of the galaxies in the proximate (and therefore contemporary) universe have violent nuclei; as astronomer George Abell of UCLA asks, does this mean that only 1 or 2 percent of galaxies ever succumb to violence, or that most galaxies—including ours—become violent 1 or 2 percent of the time?

The hypothesis that quasars are a passing phase gains support from the very fact that they shine so brightly. Quasars are so brilliant that some can be seen with an amateur's backyard telescope. If indeed they lie halfway and more across the cosmos, the amount of energy they radiate into space must be enormous. (Schmidt realized this on the day he discovered the red shift in 3C273.) And visible light is only part of the story. Many quasars are powerful radio sources as well, and, as Frank Low and Harold Johnson discovered, quasars may radiate energetically at infrared wavelengths, too. A quasar could not keep shining at that rate for anything like ten or fifteen billion years; to do so, it would have to burn more matter than is available in an entire major galaxy. But in an evolving, Big Bang cosmos, quasars need not shine for long. They might represent an early stage in the evolution of galaxies that later settled down to become "normal" galaxies of the sort we see around us today. This would square with the Schmidt-Sandage survey that indicates that quasars, rare today, predominated in the early history of the cosmos.

In recent years, the hypothesis that quasars lie at vast distances not only gained wider acceptance among astronomers, but also was being employed as a step toward conducting further investigations of deep space. Astronomers studied quasar spectra for traces of intergalactic gas clouds through which the light ought to have passed on its long way to Earth; they hoped that, by analyzing the spectra of clouds at various red shifts, they could reconstruct the chemical evolution of intergalactic clouds throughout much of the history of the universe. "Double" quasars were found that proved almost certainly to be the split images of a single quasar, its light refracted by a "gravitational lens" consisting of a massive cluster of galaxies in the foreground that bent light from the more distant quasar, just as Eddington found starlight bent by the sun in accordance with Einstein's general theory of relativity. At extreme cosmological distances, the universe may look as dappled as a windswept pond.

But if the place of quasars in the broader scheme of things was beginning to be understood, the fundamental question of how they work remained unanswered. Many quasars are not only powerful, but variable as well. Some flicker in brightness within a matter of hours. 3C446 has been observed to double in brightness, then return to normal in a single day. 3C345 will sit quietly for three months at a time, then suddenly flare up. For a quasar to vary in brightness in so short a time, its energy source must be small enough for information, traveling at no more than the speed of light, to reach all parts of it within the variation period. A quasar that pulses in, say, one day can have an energy source with a radius of no more than one light-day, or about sixteen billion miles. Were the energy source larger it would be unable to keep in synchronization, just as a marching band cannot keep time if the players scatter too widely. The message of variable quasars is that quasars not only shine more brightly than do whole "normal" galaxies, but also do so employing a kernel of energy that is tiny by galactic standards.

To some astronomers the great distance attributed to quasars, their extreme brightness and their ability to vary in brightness over short periods made the whole theory of quasars as deep-space objects look suspicious. Many of the skeptics were astronomers who earlier had been associated with the Steady State theory and who kept alert when they felt their evolving-universe adversaries had climbed out on a limb. In quasars they perceived the issue they had been looking for. A debate arose that lasted for years.

To attack the theory that quasars represented an early stage of cosmic evolution, it was necessary to challenge the assumption that the red shift in the light of quasars was caused by expansion of the universe, that quasars were far away, or "cosmological," in the parlance of the astronomers. But if the velocity was not imparted by expansion of the universe, where did it come from? James Terrell of the Los Alamos Scientific Laboratory suggested that quasars were small, relatively local objects that had been ejected from the core of our galaxy and were moving away rapidly only because the Milky Way had spewed them out. Terrell's proposal sparked some interest on the part of Fred Hoyle and Geoffrey Burbidge, who wrote a paper elaborating on it. The trouble was that if our galaxy were throwing off quasars, presumably some quasars ejected by other galaxies ought to be hurtling toward us, in which case their spectra should be shifted toward the blue instead of the red. Yet no blue-shifted quasars have been found.

Another possibility was that quasar red shifts were not caused by velocity at all but were gravitational in origin. In relativity, light is red-shifted as it climbs out of the space-time "well" of the star it is leaving. But quasar red shifts are large, and a gravitational field of the required intensity implied pronounced side effects. In a paper that even the opposition described as "overwhelming," Schmidt and Greenstein showed that the expected side effects were not present. They went on to exclude almost any nonvelocity explanation for the red shifts.

Hoyle found himself somewhat at cross currents on the issue. Just before Schmidt discovered quasar red shifts, Hoyle and William Fowler of Caltech had published a paper suggesting that the cores of powerful radio galaxies were massive collapsing "superstars." This theory stood to gain some support if quasars proved to represent an early stage in the development of galaxies. On the other hand, any suggestion that the universe was once different than it is today—that it was once full of blazing quasars—ran counter to Hoyle's Steady State theory, and though Hoyle was, by his own account, "much less interested" in the Steady State theory than he had been, he retained a certain sympathy for it. Consequently, he was disinclined to accept the view that quasars lay at cosmological distances. He called it "a fifty-fifty situation" whether quasars were really far away.

The burden of the skeptical position on quasars was taken up by a younger astronomer, Halton Arp.

The son of a commercial artist, Arp grew up in Greenwich Village and Woodstock, New York. "I had a pretty chaotic childhood, so I sort of gravitated toward a more ordered, academic existence," he said. Still, when he joined the Hale Observatories staff, some of the astronomers there came to feel that his approach was as much art as science. Arp was fascinated by the exotic and unusual, of which there is a great deal in the sky. He moved into a Moorish stone house a magician had built on a Pasadena hillside and set to work compiling his *Atlas of Peculiar Galaxies*. It bulged with photographs, some from observatory files, and others he took himself at Palomar, showing odd galaxies that caught his fancy—distorted spirals trailing like birds with broken wings, giant systems with veils of glowing gas spread between them, smoke rings, strings of beads, warped lumps like bubbling yeast. Each was home to billions of stars. Arp chose them with an artist's eye, picking what looked interesting.

Arp's approach contrasted with Sandage's. Sandage was interested in constancy in the universe: He sought distant stars and nebulae that resembled nearby stars and nebulae, galaxies and clusters of galaxies that resembled one another, in order to use them as reliable indicators of distance. Arp was after novelty.

One winter night in 1966, thunderstorms on Palomar prevented Arp, scheduled to use the 200-inch, from observing. He sat up all night in the observatory dormitory, drinking coffee, listening to rain wash over the roof and hoping the skies would clear. To pass the time he paged through a catalogue of cosmic radio sources. Radio sources were a sore point with him. He resented the radio men who had passed unpublished coordinates along to Sandage and Schmidt instead of him, enabling them to discover and explore the quasars before Arp knew what was happening. "If you got hold of one of those radio positions," he said, "all you had to do was sit in one of the big telescopes, take a photograph, take a spectrum and write a paper. It was great. But I couldn't get hold of any of those top-secret positions. I was elbowed out of the field, even though my work was right at the nexus of the study of the galaxies." As Arp looked through the list of radio sources, he thought about a letter he had received from a friend, J. L. Sérsic in Argentina. Sérsic had noticed that several compact radio sources were located in the sky near the position of galaxies from Arp's *Atlas;* he wondered if the proximity were purely coincidental. Arp wondered so also. He began comparing the celestial coordinates of radio sources in the catalogue with those of his favorite strange galaxies.

By 4 A.M. Arp had come up with several radio objects located in the sky near peculiar galaxies. "The only thing I didn't know was just what the radio sources were," he said. "In the morning I went over to the dome and looked them up in the literature. Three or four of them turned out to be quasars." In the following weeks Arp assembled a list of twenty-nine galaxies, most from his *Atlas,* whose position in the sky was adjacent to that of quasars or other radio sources. Arp suspected that the galaxies and the quasars did not just happen to lie along the same line of sight, that they were in fact physically associated. If so, the quasars could not be as far away as their red shifts indicated, but had to be at about the same distance as the galaxies. The galaxies were all relatively nearby, as intergalactic distances go. So, if Arp was right, the quasars were nearby too.

Arp soon obtained what he felt was photographic evidence of quasar-galaxy interactions. In several cases, galaxies with disturbed or unusual-looking nuclei displayed filaments of hydrogen gas that seemed to point at a nearby quasar. In others, two compact radio sources appeared along a straight line to either side of the galaxy nucleus. Arp felt these photographs constituted evidence that the quasars had been ejected from the centers of the galaxies.

Astronomers opposed to the Big Bang theory were happy to hear of Arp's work. The Big Bang had gained so much popularity that it threatened to dominate cosmology. Arp's papers attacked a central assumption of the theory, provoked questions and opened new lines of debate. Many found this healthy for the field in general. The Steady State theory might be dead, but that did not mean the Big Bang theory was correct. The universe is full of surprises; maybe the true nature of quasars was going to be one of them. In papers supporting Arp, the odds that the apparent proximity of quasars and relatively nearby galaxies he found could be coincidence were variously cited at one in ten, one in five hundred, one in a thousand and even higher.

Widely reported in the popular and semiscientific press, Arp's position generated considerable public interest. When the controversy had not cooled after five years, a debate was arranged, reminiscent of the Shapley-Curtis debate of 1920. Arp's opponent was John Bahcall, a young physicist at the Institute for Advanced Study in Princeton. Arp had taken to talking of "discordant red shifts." When he found a galaxy and a quasar that appeared to be associated, but the quasar had a much larger red shift than the galaxy, he concluded that something was wrong, or "discordant." Bahcall and his wife, Princeton University physicist Neta Bahcall, had identified a quasar with a surrounding cluster of galaxies that shared a similar red shift, a relationship they termed "concordant" in faint parody of Arp. Arp had a look at the object, and in a bristling response he charged that it was not a quasar at all. Within a few months James Gunn, a friend of the Bahcalls and, unlike them, a trained astronomer, photographed an indubitable quasar associated with a cluster of galaxies of equal red shift. This observation was an important one for the quasar orthodoxy; it supported the view that quasars really are at the distances indicated by their red shifts. Finding such missing links—clusters containing both galaxies and quasars—was difficult because galaxies were too dim to be seen clearly at distances of more than a billion light-years, while quasars didn't begin to proliferate until two billion light-years or more. Arp argued that a few such concordant examples did not matter; if he could find just one case of a low-red-shift galaxy associated with a high-red-shift quasar and prove to everyone's satisfaction that it was genuine, then the foundations of quasar research would be undermined.

Arp and Bahcall met under the auspices of the American Association for the Advancement of Science in Washington, D.C., December 30, 1972.

Arp brought along his beautiful photographs of galaxies, prints showing alleged bridges of gas and dust between objects which, by the conventional interpretation of red shifts, ought to be many millions of light-years apart. He had broadened his original thesis into a general attempt to find any combination of cosmic objects that appeared to be associated yet had differing red shifts. His position amounted to an assault on the dependability of the red-shift-distance relationship, a key to deep-space astronomy for the forty-five years since Hubble discovered it. He paraded slides before the audience: Here is a galaxy with a distended arm that points toward a quasar, here is one that seems linked to another with twice its red shift, here is a spiral clearly perturbed by the gravitational field of another that Bahcall says must be far away from it. Arp offered no real explanation of what, if not the expansion of the universe, caused quasar red shifts. His point was simply that something unexplained was going on.

Bahcall, a young man with a preoccupied manner, took the dais. He recounted the accomplishments of Hubble, Humason, Sandage and other cathedral-builders of modern astronomy, then bore down on what he saw as the central weakness of Arp's reasoning: Arp, he said, kept talking of "associations" between cosmic objects with discordant red shifts but never defined what an association was. His rules changed to fit each case. "The skies when photographed with large telescopes reveal so many individual objects on any photographic plate that one can find almost any configuration one wants if one just hunts," Bahcall said, "even stars arranged as four-leaf clovers." Bahcall charged that some of what Arp perceived as inter-

Touring America, Einstein met with some of the astronomers who were learning that the universe expands, as his general theory of relativity had predicted. *Top:* Einstein at Mt. Wilson, January 29, 1931; Hubble is at the extreme left. *Bottom:* At Yerkes Observatory, May 6, 1921. *Photos: Carnegie Institution of Washington, Mt. Wilson and Las Campanas Observatories; Yerkes Observatory*

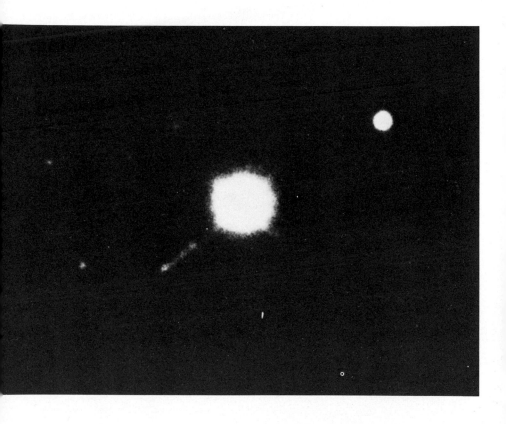

Armed with the theory of general relativity and with telescopes capable of sounding the universe to cosmologically significant distances, scientists by the mid-twentieth century were able to realize the ancient dream of investigating nature on the very largest scale. *Opposite:* A cluster of galaxies in Centaurus; the sharp round dots are stars, the fuzzy objects galaxies. *Above:* The quasar 3C273, with its protruding jet of high-velocity plasma. *Photos: Kitt Peak National Observatory*

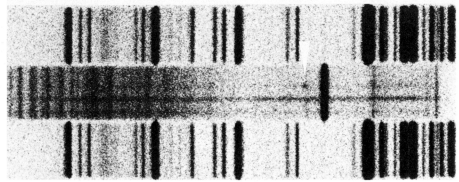

In an expanding universe, every observer finds that the more distant a given galaxy, the faster it is speeding away, and the more, therefore, its light has been shifted toward the red end of the spectrum. This, the Hubble Law, makes it possible to infer the distances of very distant objects from their red shifts alone. *Above:* A faint cluster of galaxies, receding at nearly half the speed of light, and its spectrum, displaying spectral lines displaced far toward the red. Rudolph Minkowski *(opposite)* obtained this spectrum of what were then the most distant known galaxies during his last observing run on the 200-inch telescope at Palomar, just before his retirement in 1960. The lines above and below are a comparison spectrum prepared in the observatory. *Photos: Palomar Observatory; Carnegie Institution of Washington, Mt. Wilson and Las Campanas Observatories*

The Abbé Georges Lemaître *(opposite, top,* with Einstein and physicist Robert Millikan) proposed that the universe began expanding by erupting from a state of high density, by way of a process he likened to the decay of an atom. But due to errors in distance measurement, the Hubble constant indicated an age for the universe that was less than the ages geologists assigned to the earth and astrophysicists the stars. These discrepancies prompted Thomas Gold, Herman Bondi and Fred Hoyle (seen *opposite, bottom* at a 1961 conference in Berkeley, California) to propose the Steady State theory. They suggested that the expanding universe, rather than beginning in a Big Bang, has always abided much as it does today, replenishing itself by creating hydrogen atoms out of empty space. When Walter Baade *(above),* Allan Sandage and others identified errors in Hubble's constant of expansion and brought the age of the Hubble universe into line with the estimates of the astrophysicists, the Steady State theory declined from favor. *Photos: California Institute of Technology; Thomas Gold; Carnegie Institution of Washington, Mt. Wilson and Las Campanas Observatories*

Margaret Burbidge, shown here accepting one of the many awards that graced her career in astronomy and astrophysics, worked to overcome barriers excluding women from full participation in science. In 1971 she declined the American Astronomical Society's Annie J. Cannon prize for outstanding women astronomers on grounds that "because of the small number of women in the field" it would "not be surprising if we all in our turn are selected." *Photo: University of California, San Diego*

galactic bridges were observational errors, while other "associations" resulted from a distant object falling by chance along the same line of sight as a nearby one. As for the statistical arguments by Arp and others that quasars too often appear close to peculiar galaxies to be accidental, Bahcall cautioned against statistical cases made after the fact. A photograph of Times Square may show a man from Borneo arm-in-arm with a woman from Paris, and the odds against finding such a thing may be high, but the odds of *having found* it that one time are always 100 percent, because there it is.

In rebuttal, Arp defended both his observations and his statistics. "Obviously there is always some finite chance that any one association could be accidental," he said. "If each association is considered separately, each could be dismissed on these grounds. But what is the chance that two or three or a half dozen could be accidental?"

When the debate was over, it was difficult not to be impressed by Arp's sincerity and by his love for the mysterious galaxies he studied, but it was also difficult to feel that his case had suffered anything short of demolition. Lacking a solid theory of what quasars were, he was really saying that more orthodox astronomers, the men who had "elbowed" him aside, did not know everything. That was certainly true but it did not change the fact that quasars, by the weight of the evidence, really did seem to be astonishingly far away.

To the extent that Arp did have a theory, it was that the nuclei of galaxies may explode and throw off quasars. This capitalized on the idea of exploding galaxies, something of a fad in astronomy in the 1960s. A number of galaxies with bright centers had been found, many of them strong radio objects, and in 1961 Sandage photographed one, M82, with a special filter and was astonished to see what looked like a galaxy blowing itself apart. Galactic suicide was discussed for years, with a few theorists going so far as to attempt to estimate our chances for survival should the Milky Way detonate. Then in 1976 Philip Morrison and a colleague established that M82 probably was not exploding at all. Violent galaxies may well exist, but it remained to be seen whether any were so violent as to eject large quantities of matter into space at nearly the velocity of light.*

The quasar dispute left rifts in the community of astronomers. Arp and his sympathizers were called spoilers and headline-grabbers by less impulsive colleagues, who in turn were perceived as imperious and unbending. Sandage and Arp, who had been close friends, were no longer speaking. They worked in offices fifty feet apart in Pasadena, assaying cities of stars in a language of mathematics and physics that not more than a handful of individuals fully understand, and did not talk to each other.

"I don't talk cosmology at the office anymore," Arp said one dusk, at home, as he watched the sunset. "It's a

* The M82 episode marked one of two occasions when the normally prudent Sandage erred. The other came in 1965 when he wrote a paper titled "The Existence of a Major New Constituent of the Universe: The Quasi-Stellar Galaxies." Using a two-color survey technique he had developed with Ryle, Sandage found many quasars that did not radiate radio noise. The essence of the paper was sound, in that radio-quiet quasars do outnumber radio-prominent ones, but Sandage overlooked the fact that many of the objects he took for "quasi-stellar galaxies" had already been studied by Caltech astronomer Fritz Zwicky, among others. Zwicky found that they were compact galaxies that probably had a place in quasar-to-galaxy evolution but hardly deserved the label "major new constituent of the universe," since Zwicky already had observed them. Sandage was apologetic. "I rushed in too quick," he said.

very touchy subject. The people committed to the other side take it pretty seriously. Part of it is ego. I have a friend who says, 'Any man who's egotistical enough to think he can figure out the whole universe is pretty damn egotistical.' The other part is that the success or failure of your theories gets associated with the success or failure of your scientific career, and most people here have their whole lives wrapped up in their careers.

"You know, a lot of us, growing up, wanted to be famous scientists. At Caltech we struggled to learn very complicated things. Our lives, our success, depended upon whether we could pass those exams, regurgitate stuff like Hubble's red-shift-distance law. Now, in later years, having learned it has become almost a part of our security. People take it seriously and if somebody questions its validity, their psychological security gets shook up.

"It's so competitive these days. Pressure on the telescope, people scrambling for time—that's the gut issue, who gets what amount of time on the 200-inch. Every time you go down to Palomar you feel you have to make an astounding discovery or you're in peril. I think graduate school failed to get across to everyone the idea that it would be all right to have a pluralistic society, astronomers working on different things and coming up with different results. I think that failure is unfortunate.

"Sooner or later we will find out what quasars really are, but whether we find out how to communicate with and trust one another, well, that seems to me much more important."

Arp and Sandage began speaking again within a few years and soon had regained their old friendship. One morning they agreed to sit down together in Arp's office to discuss the interpretation of quasar red shifts. Their appearance, like their science, was a study in contrast—Arp in a gray pinstripe suit as meticulously tailored as a diplomat's, Sandage with his large deep eyes and runaway hair resembling an Old Testament prophet. The atmosphere was jovial.

"You're the master of the conventional view, Al, and there's a lot of truth in the conventional picture," Arp said. "It all fits together so beautifully. But there are observational discrepancies that must be faced."

"I don't understand how you can accept *both* the conventional picture and your own view, which lies outside anything that can be called theory," Sandage replied. "Sometimes you rely on one, yet at other times on the other picture. . . . What do you conclude from the associations you see between galaxies and quasars?"

"I think the quasars were ejected from the galaxies," Arp said.

"What, then, produced the quasar red shift?"

"I don't know. That's how we make progress in science, by confronting what we don't know. . . ."

"What is the mechanism responsible for the mystical red shift you see as superimposed upon the Doppler red shift [of the expanding universe]?"

"I make no prediction as to its source," Arp replied. "But I do predict its existence."

"Then that's not a prediction."

"I want to get this off my chest, Al," Arp said. "I don't care what the theory says I can observe. It's the observation, not the theory, that defines the reality."

"Upon what basis, then, is the prediction made?" Sandage asked.

"It's empirical. It's based upon my experience."

"OK. So something happened in the past, nature did it once so it will do it again. It's a continuity argument. But that is not, in the sense of the philosophy of science, a prediction. . . . In your *modus operandi* you are a pure Baconian. You induce from below without any preconceived notions. That's fine, except that most observations include false clues. Science proceeds both ways—by induction from below *and* by deduction from hypotheses above. Prediction comes only from deduction. A pure empiricist, I believe, will be led astray more often than not."

"But Al, imagine yourself back five hundred years in human history. . . ."

"We're not there," Sandage said in the mock-ominous voice of an early radio announcer.

Arp laughed. "How could you have worked purely from deduction back then?" he asked.

"You couldn't have."

"Then are you saying that we're at the golden millennium?" asked Arp. "That now we know everything? That only today do we know enough to go ahead and work deductively?"

"You're black-and-whiting it," Sandage said. "The theories today have had sufficient success that we can proceed deductively with greater confidence than we could have centuries ago."

"Do you think that five hundred or a thousand years from now we will still believe the same theories we do today?" asked Arp.

"I don't know. But I object to your saying that we don't know anything about anything. . . . We couldn't have gotten modern physics from Bacon. Pure induction alone wouldn't have given us modern physics, because all empirical data include false clues."

"That's true, but we have to work with what we have," Arp said.

"It's theory that causes science to progress," Sandage remarked as the conversation progressed. "I would feel skeptical if someone were to tell me, 'I saw an apple fall upward,' because it is contrary to an extremely well-founded theory. In that case I would believe that the 'observation' was *wrong.*"

"I'm not saying that I saw apples fall upward."

"I think you are. . . ."

"I disagree."

"There's an aesthetic difference between us," said Sandage. "Your style and appreciation of science differ from mine."

"One of us is right and one of us is wrong," Arp replied.

"We're all wrong to some degree," Sandage said. "But the universe is the way it is, independent of our views. The universe doesn't care what I think or what you think. That's the beauty of science. Its beauty and truth are independent of the beliefs of the scientists. . . ."

The two talked on, laughing frequently and disagreeing about almost everything, as the late-morning sun climbed across the window of Arp's office. "In the long term the truth comes out," Sandage said. "One should never imagine that ultimate truth has been uncovered."

"Amen," said Arp.

In 1981, the committee allocating Palomar observing time informed Arp by memo that he should not expect to be granted further access to the 200-inch telescope if he persisted in using it to investigate associations between quasars and galaxies.* Arp's years of work on discordant red shifts had produced few verifiable predictions, the memo emphasized, and his views had failed to win the support of more than a small minority of astronomers. The committee members' tone was impatient, almost intemperate; they had made an exception of Arp for years, were tired of doing so and now were going to put a stop to it.

Sandage came to Arp's defense. He said he felt that Arp's work had some value, especially now that Arp was at last prepared to make an in-depth survey of a swatch of sky large enough for meaningful statistical conclusions to be drawn about the distributions of galaxies and quasars. And Sandage didn't care for the memo. That wasn't the way to treat a senior astronomer, he said. Unswayed, the committee in 1982 sharply reduced the time allocated to Arp on the 200-inch telescope.

By and large, Sandage socialized little in the astronomical community, attending scientific conferences rarely. At one meeting, Geoffrey Burbidge, in a jovial mood, drew a horizontal line on a blackboard and with the help of the audience charted astronomers along the line according to whether, and how strongly, they felt quasars were denizens of early cosmic history. Arp was awarded the extreme left wing. Schmidt, who was devoting almost all his time to quasars on the assumption that they were far away, was placed at the far right. Margaret Burbidge asked to be put slightly left of center. Bahcall went to the right, just shy of

---

* Caltech and the Hale Observatories had sundered the alliance Hale had forged, under which they jointly administered Palomar and Mt. Wilson. Caltech now took full command of Palomar. The Hale Observatories office, funded as before by the Carnegie Institution but renamed the Mt. Wilson and Las Campanas Observatories, operated Mt. Wilson, as well as a newer facility beneath the lucid skies of the Chilean Andes.

Schmidt. Where, someone asked, do we put Sandage? Burbidge thought for a moment and then, to general laughter, wrote in Sandage's name on the extreme right but high above the others. Sandage was above it all. He had not come to the conference. He was on the mountain-top, trying, as he sometimes put it, "to rewrite the book of Genesis."

In his office, where a portrait of Eddington hung on the wall, and every level surface was stacked feet deep with photographs, technical papers and computer print-outs, Sandage said, "It doesn't matter whether there is a controversy over quasars. It doesn't matter what anyone says. The ultimate judge is nature. The majesty and excite-ment of the thing is just that, that it is independent of man, *and it is out there*! The galaxies will go on regardless of the councils of the scientists. There is an *ultimate truth*!"

He pushed a red-shift-distance diagram across the desk.

"This," he whispered, "this is what nature *says*!"

9

THE FATE OF THE UNIVERSE

The explorations of space end on a note of uncertainty. . . . We measure shadows . . . we search among ghostly errors of measurement. . . .
— EDWIN HUBBLE, 1935

SANDAGE DESCRIBED his life's work as "a search for two numbers." One was the rate at which the universe is expanding, known as the Hubble constant, designated H_0. The other was the deceleration parameter, q_0. It represented to what degree the expansion was slowing down. If these two numbers were known they could establish a great many things about the dynamics of the universe, including, perhaps, whether it will expand forever.

Revisions in the Hubble constant over the years were so radical as to be something of an embarrassment; one astronomer wrote that they played "in the history of our subject much the same part as the crash of 1929 in the history of Wall Street." With each revision the universe was perceived to be larger and older, and therefore to be expanding at a more gradual rate, than had been thought before. When Sandage was a graduate student at Caltech, the Hubble constant was being quoted at about a hundred miles per second per million light-years. That is, for every million light-years farther out one looked, one would observe galaxies receding at an additional hundred miles per second velocity. Baade's work on Cepheid variable stars cut this figure in half; in 1956, astronomers on three continents reduced it further; Sandage scaled it down still more in the 1950s and again in the 1970s, until the "constant" wound up at 10 percent of its original value. According to Sandage's 1976 figures, galaxies recede at ten miles per second per million light-years distance. Sandage claimed a margin of error of under 15 percent. "It's a real number now," he said. This would make the age of the universe (meaning time since expansion began) eighteen billion years, give or take some three billion.

Determining the rate at which the universe expands consumed much of Sandage's time for a period of twenty years. He had to obtain thousands of red shifts and to find accurate distances for each galaxy by some means independent of red shift as well. The first chore was not especially difficult, but the second was. To determine galaxy distances, Sandage elaborated upon the stairstep procedure of Shapley and Hubble. First he rechecked distances to relatively nearby galaxies by isolating Cepheid variables and other recognizable types of stars in them; the absolute magnitude of these stars had been determined by studying them in our own galaxy, and by comparing their absolute magnitude with their apparent brightness in the sky, the distances of other galaxies could be estimated. This first step took Sandage out to about five million light-years, the limit to which the 200-inch telescope could detect Cepheid variables. To go farther, he examined certain very bright components of the nearby galaxies, primarily giant blue stars and knots of glowing hydrogen gas; when he knew the typical intrinsic brightness (and, in the case of the gas knots, the size) of these features with confidence, he then could estimate the distance to any galaxy in which they could be photographed. This step was useful out to thirty million light-years.

Beyond that, Sandage shifted from using stars in galaxies to using entire galaxies as his "standard candles." This procedure was made feasible by the remarkable fact, discovered by Hubble, that clusters of galaxies resemble one another. Each cluster typically contains one giant elliptical galaxy or one dominant spiral, and a type specimen in one cluster is likely to be of roughly the same brightness as its counterpart in another cluster. By studying many clusters of galaxies, Sandage was able to learn to what extent this grand-scale display of the uniformity of nature could be trusted. He then could employ standard galaxies in clusters of galaxies to measure distance, as once he had employed stars. He penetrated well beyond three hundred million light-years in this way, sampling enough of the universe to feel assured that he was measuring expansion unfettered by local deviations. (The expansion of the universe does not occur within clusters of galaxies, which are bound together gravitationally, but out between clusters—and since there is reason to suspect that the clusters themselves are organized into metagalaxies, it is necessary to observe a large number of galaxies to be sure that the observed expansion rate is really typical of the cosmos and has not been slowed by the proximity of clusters.) A number of other astronomers contributed to this work, although Sandage dominated the field. A variety of other distance-measuring techniques were employed, including apparent angular diameters of galaxies and their radio profiles. The idea was to try as many independent methods as could be found in order to reduce the margin of error, to mount "enough lines to attack to overdetermine the parameters," as Peebles at Princeton put it.

The result, when red shifts were plotted against distance, was the straight line of a uniformly expanding universe.

Gathering the data was dry toil that Sandage commenced as a young man and concluded in middle age. It held little glamor for him. Most mornings he had to force himself to sit at his desk and resume work. "Self-discipline," he liked to say, "is the highest human virtue."

The second number, the deceleration parameter, remained.

Standard Big Bang theory assumes that the expansion of the universe must slow down as time passes, since the energy of expansion was imparted all at once, in the beginning, and the mutual gravitational attraction among the galaxies should have been acting as a brake on it ever since. The question is, at what *rate* is the expansion of the universe slowing? If the deceleration is sufficient to halt expansion altogether one day, collapse of the universe should follow; if so, the universe is "closed," in the sense that both its past and its future are finite in time. If deceleration is less than the critical value, expansion will continue forever and the universe is "open," its future infinite in time. A closed universe may be called spherical or elliptical in the language of the noneuclidean geometries. An open cosmos is hyperbolic. The models, of course, are four-dimensional—one dimension of time plus the three dimensions of space—which is why the spacial geometry of the universe presumably can be deduced if we know its fate.

A deceleration rate sufficient to close the universe will show up as a departure upward from the straight line on a chart of the Hubble velocity-distance relation. The departure appears because at high lookback times we are seeing a period in cosmic history when the universe expanded more rapidly than it does today; the difference between the velocity of expansion then and now yields the rate of deceleration. The universe, however, is vast, and the Hubble relationship must be tested to great distances before any deceleration rate is likely to become evident. Spectra of galaxies billions of light-years away are difficult to record and analyze, and independent checks on distance are fraught with error. From the outset Sandage remained pessimistic about ever being able to establish the deceleration parameter by studying galaxies with earthbound telescopes. A telescope in space, above the Earth's turbulent atmosphere, could do the job, but a space telescope was still in the future.

In 1972 Sandage attempted to extract the deceleration parameter from the data he and three other astronomers had accumulated in the course of refining their value for the Hubble constant. Since the effect of the deceleration parameter was expected to be subtle—smaller, in fact, than Sandage's margin of error—he enlisted a computer program designed to eliminate "noise" from the data and seek out an underlying trend. The results suggested a closed universe but were highly uncertain. "No decision is yet possible . . ." Sandage wrote, "as to whether the universe is open or closed."

An alternative was to look for the deceleration rate not in gross numbers at great distance, but in subtle trends closer to home. Clear-cut expansion of the universe, or "pure Hubble flow" as it was sometimes called, begins about three hundred million light-years away, where local gravity is no longer an important inhibition. Working with Gustav Tammann of Switzerland and Eduardo Hardy of Chile, Sandage studied the recession rate of galaxies lying just beyond that distance. Thirty clusters of galaxies were sampled, their red shifts and distances measured with the greatest possible accuracy. All fell almost perfectly on the flat Hubble line. There was no perceptible tendency toward deceleration.

That the universe expands homogeneously in all directions was well established, but Sandage had not expected expansion, even locally, to be *that* homogeneous. The thirty clusters showed *no* concerted variation from uniform expansion. Several giant aggregations of galaxies, including the rich Virgo, Coma and Ursa Major clusters, lie not far, intergalactically speaking, from the clusters Sandage studied, and he expected their gravity to have slowed expansion locally. Instead, all the clusters appeared to be flying apart from one another, and from ours, at an unfettered rate. This indicated that the universe was expanding with almost unchecked vitality, that the deceleration rate must be small.

More evidence to the same effect came when improved studies of the stars in globular clusters, previously thought to be on the order of eleven billion years old, indicated they were more nearly fourteen billion years old. The older the cosmos, the more powerful the Big Bang must have been to result in the rate of expansion we see today, and so the less likely it is that deceleration is enough to close the universe. The new globular cluster age estimates also fit fairly well with Sandage's value of eighteen billion years since the Big Bang, a number obtained by assuming the universe has been expanding without significant deceleration since genesis.

In 1974 four younger astronomers—J. Richard Gott III and James Gunn of Caltech, and David N. Schramm and Beatrice Tinsley of the University of Texas—wrote a long paper summarizing evidence that the universe is open and will expand forever. In contrast to the lofty, solitary tone of Hubble and Sandage's publications, their paper was the work of a team. It drew on the work of sixty-four astronomers, and it had a wry, ironic tone compared to that of the older cosmological cathedral-builders. It began with a quotation from Lucretius: ". . . In all dimensions alike, on this side or that, upward or downward through the universe, there is no end." It approached the question of the fate of the universe in terms of matter: Does the cosmos have enough matter in it to generate the gravity needed to stop expansion? Evidence from a variety of sources suggested it does not.

If the estimated mass of all galaxies is added together, the Gott group noted, the total is too little by a factor of ten to close the universe. One could propose that there is a lot of dust and gas out between galaxies, but our ability to see clearly for enormous distances—to over ten billion light-years, if the quasar red shifts are cosmological—suggests that intergalactic space is clear. The Gott group considered several suggestions about where large amounts of matter might be hidden and rejected nearly all of them. Where could the "missing matter" be? "Loopholes in [our] reasoning may exist," they wrote, "but if so, they are primordial and invisible, or perhaps just black." Black, that is, as a black hole.

Insular, excommunicant, perpetually hungry, a black hole is the perfect place to hide anything. It is a well in space-time so steep that nothing, not even light, can climb out of it. A typical black hole is a confluence of the great and the small, in circumference smaller than the city of Boston, in diameter infinite—you can feed a tape measure in forever and the black hole will simply eat it. It will eat stars and planets, flowers or kings with equanimity. Each, falling in, departs our universe. The black hole will tell us nothing about its victims. It will tell us, in fact, nothing about itself except its mass, electrical charge and angular momentum. "A black hole has no hair," as John Wheeler liked to say.

That these baleful sinks exist was a prediction of relativity. Karl Schwarzschild, one of the few who realized, prior to Einstein, that the universe might be viewed in terms of noneuclidean geometry, went to work soon after general relativity theory was published in 1916, investigating the geometry of space-time in the vicinity of massive objects, which practically speaking meant stars. His calculations revealed that for a star of any given mass there is a "Schwarzschild radius" such that if the star is compressed to a size smaller than that radius, it will collapse to so high a density as to pinch itself off from surrounding space-time. The result is a kind of hole in space. Things can fall in but nothing gets out.

The Schwarzchild radius became part of theoretical astrophysics, but whether a real star ever suffered such a fate was another matter. Densely packed white dwarf stars were known to exist and were believed to result when stars burned out their nuclear fuel and collapsed, the outward pressure of their light and heat no longer able to balance the grip of their own gravity. White dwarfs were fantastic objects—a cubic inch of their surface material could weigh tons—and the possibility that stars might collapse even further was not taken very seriously at first. Then in 1930 a young Indian student, Subrahamanyan Chandrasekhar, calculated that if a dying star were much more massive than the sun, it would collapse right through the white-dwarf stage to become something even more compressed.

Chandrasekhar was studying under R. H. Fowler at Cambridge at the time, and Fowler brought his work to

Eddington's attention. Eddington saw that if the calculations were correct, a sufficiently massive collapsing star could become so compressed that it sucked in its own light and cut itself off from the rest of the universe. This struck Eddington as ridiculous, and he felt he had disproved Chandrasekhar's theory by reducing it to absurdity.

In America the nuclear physicist J. Robert Oppenheimer, urged on by, among others, George Gamow, looked into the physics of gravitational collapse, and with several students published two important papers in 1939. The first theorized that stars could collapse through the white-dwarf stage and become "neutron stars," objects ten thousand times denser than white dwarfs that resembled nothing so much as giant atomic nuclei. The term was borrowed from Baade and Zwicky, who five years earlier had suggested that exploding stars might leave neutron stars behind as a sort of ash.* The second Oppenheimer paper discussed the possibility of a star crushing itself still further, beyond the observable limit. "When all thermonuclear sources of energy are exhausted a sufficiently heavy star will collapse," the paper began. "Unless fission due to rotation, the radiation of mass, or the blowing off of mass by radiation, reduce the star's mass to the order of that of the sun, this contraction will continue indefinitely. . . . Light from the surface of the star is progressively reddened, and can escape over a progressively narrower range of angles." As the gravitational red shift approaches infinity, the star disappears. The result is a black hole.

Interest in black holes reawakened in the 1960s with

---

* They were vindicated thirty years later with the discovery of a pulsar, or radio-bright neutron star, in the Crab Nebula, where Chinese astronomers had recorded a supernova in the eleventh century.

the discovery of quasars, the great energy of which might be explained by gravitational collapse, and of pulsars, which were shown almost certainly to be neutron stars. If there are things in space strange as neutron stars, one reasoned, why not black holes? Theoretical models showed that while finding solitary black holes in space by optical means might be impossible, they could be detected if they were part of double-star systems: the black hole would suck surface gas off its companion star, and the gas, whirling down into oblivion, would emit a flood of X rays.

Their exotic name aside, X rays are just another part of the electromagnetic spectrum, occupying the zone between ultraviolet light and gamma rays. X rays cannot be observed reliably from Earth because the atmosphere tends to absorb them. An Italian-American X-ray satellite, Uhuru, launched in 1970, eventually located over a hundred cosmic X-ray sources, of which some were other galaxies, some neutron stars and one of which, Cygnus X-1, was possibly a black hole. The issue remained in doubt, but several researchers eventually pronounced themselves almost certain they had found a genuine black hole. Cygnus X-1 appeared to be feeding off its companion, the supergiant star HDE 226868, in a gravitational web so tight that the two whipped around their common center of gravity once every 5.6 days. If that model were correct, the system would be quite a sight—the tormented supergiant pouring glowing gas across space and into an invisible space-time vortex—but interstellar cosmonauts might be forgiven for passing up a close look. The crew of a spaceship falling into a black hole would experience, in their last moments, a growing red shift of starlight resembling the death of the universe. But the death would, of course, be their own.

Although scientists were to argue for years whether it and other candidates located thereafter really were black holes, public response to discovery of Cygnus X-1 was enthusiastic. Science-fiction writers made room for black holes in stories, poets heralded them, a few painters accepted what may be the ultimate technical challenge and attempted to paint their portrait. California physicists considered the feasibility of black-hole power plants. Soviet theorists proposed that a massive, rotating black hole might constitute a bridge between two distant parts of the cosmos, one that a daring cosmonaut might navigate. In Texas two physicists proposed, deadpan, that a tiny black hole caused the Siberian forest explosion of June 30, 1908, when a brilliant fireball of uncertain origin slammed to earth near Tunguska, a hundred miles north of the Sea of Okhotsk, leveling three hundred square miles of trees. They imagined that the black hole continued through the Earth and emerged from the North Altantic between Newfoundland and the Azores. They urged that ships' logs of the period be searched for sightings of the reappearance.

As Kip Thorne of Caltech wrote, "Of all the conceptions of the human mind from unicorns to gargoyles to the hydrogen bomb, perhaps the most fantastic is the black hole."

To cosmologists, the significance of black holes lay in their potential role as lockers for hidden matter. All the known matter in the cosmos is not nearly enough to halt the expansion of the universe. Unless ten times more matter can be found, the universe is open. Could 90 percent of the stuff of the universe be locked away in black holes?

If all black holes were formed in the collapse of stars, probably not. Supposing that massive stars have been dying at a constant rate since the galaxies formed, a spiral system the size of the Milky Way ought today to contain about a hundred million black holes. To avoid erring on the low side, let us say that stellar collapse was once more common than it is today, and boost the estimate by a factor of ten, to one billion. Assume, again on the high side, that the average black hole mass is ten times the sun's. (A more likely estimate is half that.) The result is to add ten billion solar masses to our galaxy. That is a lot of mass, but it makes the galaxy heavier by less than 5 percent. We can go on to assume that every galaxy once had a quasar at its center and that the quasar remnant remains as a black hole. Theoretically this giant black hole could have a mass of up to one billion times the sun's. Even if we double that, we are still a long way from finding enough mass to stop the expansion of the universe.

Gott and Gunn considered these and other "missing mass" theories, discarded them, then concluded, by studying the gravitational interaction of galaxies in clusters, that galaxies simply do not act as if they are ten times more massive than currently estimated. *"Galaxies themselves cannot close the universe,"* the Gott group wrote (their italics). If black holes harbor most of the mass of the cosmos, they must not be in galaxies.

The Gott group concluded that the universe probably, though by no means certainly, is open. "A clear verdict is unfortunately not yet in, but the mood of the jury is perhaps becoming perceptible," they wrote. Sandage, who had leaned toward a closed universe, shifted toward the open view, prompted by the globular-cluster star ages and his own studies of expansion rates nearby. "The conclusion that the universe will expand forever seems inevitable," he wrote in 1976.

If so, the universe seems destined for heat death. Eventually all the stars will go out and the material for making new ones will have been exhausted. The cosmos will be an entropic duchy, carting black galaxies outward in a black cosmos forever.

If not, expansion will one day stop, contraction follow and a new cosmic fireball be born. Present evidence weighs against this but is not conclusive. As John Wheeler said, "It's too soon yet to say what the story is. The excitement is just beginning. Nobody should throw in his hand at this point."

Whether the cosmos ends in fire or ice—in the words of the poem Robert Frost wrote after Harlow Shapley talked with him about cosmology—it will end, in the sense that it will evolve into something we would not recognize and could not live in. Some find this depressing. Bertrand Russell regarded it as tragic that "all the noonday brightness of human genius . . . the whole temple of Man's achievement must inevitably be buried beneath the debris of a universe in ruins. . . ."

But later he cheered up a bit, as might we all. We have little more personal stake in a cosmic transfiguration a hundred billion years in the future than do sunflowers or bottleflies; snap your fingers twice and you will have consumed a greater fraction of your life than all human history is to such a span. Life is not edifice but process. We owe our lives to universal processes, in that identity of cosmic and earthly destiny glimpsed by George Ellery Hale, and as invited guests we might do better to learn about them than to complain about them. If the prospect of a dying universe causes us anguish, it does so only because we can forecast it, and we have as yet not the slightest idea why such forecasts are possible. A few figures scrawled on a piece of paper can describe the rate the universe expands, reveal what goes on inside a star or predict where the planet Neptune will be on New Year's Day in the year A.D. 25,000. Why? Why should nature, whether hostile or benign, be in any way intelligible to us? All the mysteries of science are but palace guards to that mystery.

10

EXPANDING UNIVERSE OF THE MIND

What led me to my science and from my youth filled me with enthusiasm, is the fact—not at all self-evident— that our laws of thinking conform with the lawfulness in the passage of impressions which we receive from the other world, thus making it possible for man to gain information about that lawfulness by mere thinking.

—MAX PLANCK

AS I HAVE BEEN TELLING the story, we in this century first saw the universe for what it is. When the century began we knew little of the cosmos beyond our immediate stellar neighborhood. By the time it was three-quarters over we knew that galaxies exist, that stars are born and die, that the universe expands and was born in an eruption whose rumbling echo we still can detect. The age of the expanding universe has been determined and telescopes have plumbed to the edge of cosmic history. This rich fund of knowledge was acquired by a relatively small community of scientists working with modest resources over only a few decades. It is a dramatic story. Is it true?

Reports of scientific discovery, this book included, I'm afraid, tend to give a spurious impression of great progress recently attained. They suggest that humankind labored in ignorance for centuries until a few years ago, when the light of wisdom dawned. I think this tendency comes about because discoveries, by their nature, make good stories, while enduring bafflement does not; the storyteller concentrates on what has been learned and ignores what has resisted comprehension. In any event, it is a distortion. Now as before, the unknown dwarfs our little precincts of the known. Newton was not merely being modest when he said, "To myself I seem to have been only like a boy playing on the seashore, and diverting myself in now and then finding a smoother pebble or a prettier shell than ordinary, whilst the great ocean of truth lay all undiscovered before me."

The history of science resembles less an efficient machine than a rummage drawer of tools, most of them broken and discarded. In the late nineteenth century, physics was widely thought of as a closing book. Gravitation had been accounted for by Newton, heat by Maxwell and Boltzmann, electromagnetism by Maxwell and Faraday, and with such apparent success that students were cautioned to expect to make no breakthroughs in their careers but to content themselves with the thought that physics henceforth would consist of little more than refining existing "laws." Then in 1900 Max Planck of Berlin discovered the quantum principle, and science was transformed. Almost every "law" was altered, diminished or discarded. Alfred North Whitehead, born in 1861, said in the 1930s, "Nothing, absolutely nothing was left that had not been challenged, if not shaken; not a single major concept. This I consider to have been one of the supreme facts of my experience."

To imagine that cosmology is immune from similar revolution is to invite a shock.

The Big Bang cosmology—by which I mean not any single theory but rather all violent-genesis models built on general relativity—accounted for the expansion of the universe and for the quasar cutoff point, and it accurately predicted the discovery of the cosmic background radiation. Those were remarkable achievements, and upon the basis of them the theory must be termed a success. But to say that a theory is successful is not the same thing as to say it is true. The cosmology of Claudius Ptolemy survived for thirteen hundred years, and, with various modifications, predicted, with reasonable accuracy, the motions of the planets in the sky all that time. Yet its central assumption, that the sun and planets circle the Earth, was false. The sun-centered cosmologies of Copernicus and Kepler somewhat improved the accuracy with which theory fit observation; the accuracy was increased further with Newton's and, in turn, Einstein's theories of gravitation. But all these theories were human inventions, and none should be confused with the phenomena they sought to explain. It seems unlikely that "gravity" existed in nature and that Newton "discovered" it, or that nature contained a "space-time continuum" before Einstein discarded gravity and erected the continuum in its place. Our theories are not "laws" that nature "obeys." They imitate the sky; they did not fall from it. So it aids clear thinking to keep in mind that the Big Bang theory is not the universe, but only a preliminary, incomplete description of how, in a few very limited ways, the universe may behave.

What we do not know is an ocean; I will call attention to a few recognizable inlets and estuaries, and those briefly. I cannot go farther, in part because beyond the light of our small knowledge, ignorance compounds itself—we do not know what we do not know. I think this is what Fred Hoyle had in mind when, at a conference where a speaker remarked that "there are questions we cannot answer," Hoyle whispered to the physicist Philip Morrison that in science the answers are not important, the questions are. The scientist who asks the right question reconnoiters a new patch of the unknown, and may, with luck, bring it within the constricted but expanding boundaries of the known.

The recent history of science offers a vivid lesson in question-asking. It is the advent of quantum physics, the revolution Whitehead spoke of, which changed not only science but, to some degree, the foundations of science, its deep presuppositions. By telling a little of this story and giving a few examples of how it has affected modern cosmology, perhaps I can hint at something of the form of the frontiers of science.

Planck developed quantum theory in response to the "blackbody" problem that plagued experimenters. The experiment in question consisted of observing the radiation of heated gas in a neutral ("black") enclosure. By hooking up a spectrometer and temperature-measuring device, an experimenter could chart the intensity with which the gas radiated energy at various wavelengths. When this experiment was performed, it gave results that conflicted with the predictions of classical theory. Each gas tested yielded a curve—wavelength against temperature—that had a characteristic "blackbody" contour, with its peak clearly related to the temperature of the gas, but the shape of the curve and location of the peak were not those predicted by existing physics. Attempts to reconcile theory and observation failed. Ultimately Planck realized that only by breaking with the scientific tradition in which he had been raised could he account for the blackbody results. After months of what he called the most intense work of his career, he derived the formula since called "Planck's law." It has accurately predicted the behavior of radiated energy in laboratory jars, in the sun and stars and, apparently, in the universe at large—the cosmic background radiation discovered by Penzias and Wilson conforms to a Planck curve.

Planck saw that atoms radiate energy not in a smooth continuum, as had been assumed, but in discrete packets he called "quanta" (after the Latin "quantus," for "how much"). Quantum theory holds that nature acts, in a sense, like a bank teller who can pay out a penny or two pennies but not a penny and a half. "There is no way out,"

Planck told his students. "We have to become accustomed to the quantum theory, and we shall see that it will penetrate into more and more fields of our physics."

Nineteenth-century science viewed matter and energy as two different worlds, each with its own set of "laws." Matter was said to be composed of particles, energy of waves. Planck's theory violated this well-established distinction by portraying energy as composed of "particles," the quanta. Within a generation, the Parisian physicist Louis de Broglie showed that matter could be viewed as waves, just as energy could look like particles. Indeed, De Broglie said in his Nobel Prize address in 1929, "to describe the properties of matter, as well as those of light, we must employ waves and corpuscles simultaneously."

The general reaction among researchers was to say, well, fine, particle theory and wave theory can both be applied to matter or energy, but ultimately nature must be one or the other. Which is it *really?* Particles or waves? Much of the theorizing and experimentation of early twentieth-century physics revolved around this question.

Two schools emerged and briefly coexisted. They represented the old wave-particle question in new clothing. Physicists who were intellectually more comfortable with a particle universe proved that the waves could be interpreted as probabilities: The ultimate reality was particles, in their view, and the waves reflected the probability of finding particles at each given point along, say, a ray of light. Einstein pioneered this position, in an early paper explaining light as composed of quanta he called photons. Those favoring wave theory, notably the Viennese physicist Erwin Schrödinger, demonstrated with equal plausibility that *waves* were what counted, that particles were an illusion. Each side delighted in inventing experiments that made the other's position seem dubious. At this both sides were equally successful. Schrödinger, followed by Dirac, then demonstrated that the two schools were mathematically equivalent, that they amounted to different ways of saying the same thing.

But what was nature *really?* A swarm of particles? An ocean of waves?

The twenty-six-year-old physicist Werner Heisenberg, studying under Niels Bohr in Copenhagen, concluded that the paradox of waves and particles could be resolved only by taking the role of the observer, the physicist, into account. This step led Heisenberg to establish the uncertainty principle, and it is fair to say that science has not been the same since.

As Heisenberg recalled it, he was set on the path to the uncertainty principle by a remark Einstein had made in a private talk the previous year. Einstein was disturbed by what he thought to be Heisenberg's almost ruthless refusal to concern himself with elements of the physical world other than those that could be observed. Specifically, Einstein was concerned about a paper by the young Heisenberg that failed to accept the notion that electrons orbit the nucleus of atoms. Heisenberg refused to talk about electron orbits because no electron had ever been observed in orbit, and in all likelihood none ever could be.

"But surely you don't believe," Einstein said, as Heisenberg remembered the conversation, "that none but observable magnitudes must go into a physical theory?"

"Isn't that precisely what you have done with relativity?" Heisenberg asked. "After all, you did stress the fact that it is impermissible to speak of absolute time, simply because absolute time cannot be observed; that only clock readings, be it in the moving reference system or the system at rest, are relevant to the determination of time."

"Possibly I did use this kind of reasoning," Einstein said, "but it is nonsense all the same. . . . It is quite wrong to try founding a theory on observable magnitudes alone. In reality the very opposite happens. It is the theory that decides what we can observe. . . ."

Late one night in February 1927, Heisenberg walked in Copenhagen's Faelled Park, thinking about Einstein's remark, "It is the theory that decides what we can observe." That certainly seemed to describe the question of wave theory versus particle theory; one could see the subatomic world as waves or particles, whichever one looked for. What if the question of which was "real" could *never* be finally decided? What if the sort of things physicists observed in the subatomic realm was determined less by "reality" than by the methods they used to observe? Heisenberg knew, for example, that one could not observe an electron in an atom without knocking it out of the atom: The radiation needed to "illuminate" the electron for observation—in practice, gamma rays were employed for this purpose—would knock it out of orbit. This had been the bone of contention with Einstein, who thought of the orbit as "real" even though an electron could not be observed in orbit. Heisenberg wondered whether this limitation of observation might be the real issue. Perhaps we can never observe the "real" world on such a small scale. If we cannot, Heisenberg felt, then speculation about what that world was "really" like was not, in this situation at least, a matter that science could decide.

Back in his flat, Heisenberg attempted to determine just how much uncertainty was built into the situation. If a physicist wants to know exactly where a "particle" is at a given moment, he must interfere with it—wallop it with gamma rays or some other form of radiation—to find out, and in the process he will change its behavior, knocking it onto a new path. The price of knowing its position is that he must relinquish knowledge of how it might have behaved if he had let it alone. If, on the other hand, he wants to know the particle's behavior in space and time, how fast it is going and in what direction—and it is this behavior of great numbers of "particles" that generates "waves"—he must wait and see; a track in a cloud chamber or an oscilloscope will tell him where the particle went, after the fact. The price of that observation is that the physicist cannot know exactly where the particle was at a given moment. The trade-off, Heisenberg realized, is fundamental. The subatomic world is grainy with inexactitude, is built of little uncertainty "boxes" we macroscopic observers can *never* peer into. We can squeeze the box one way to make things look like particles, or the other way to see waves, according to how we set up each experiment, but we cannot reduce the area inside the box. When Heisenberg computed the "amount" of uncertainty inside the box, he founded it defined by Planck's constant.

When Bohr returned to Copenhagen—he had been away on a skiing trip—and was presented with Heisenberg's new theory, he went to work extrapolating it. The result was his "principle of complementarity." Bohr demonstrated that the uncertainty principle was implied by fundamental laws of conservation of energy and momentum. Wave and particle models were complementary aspects of the same reality. The "real" nature of the microscopic world, it appeared to Heisenberg and Bohr and has appeared ever since, cannot be captured in a microscopic model.

I have attempted this shamelessly oversimplified account of early quantum physics in order to suggest that something of deep importance was going on in science. Science was beginning to leave behind its need to visualize, and with it was discarding some other trappings of traditional science, including the doctrines of force and causation.

Scientific theories must be logical, must be expressible in terms of mathematics, the most rigorous logical system known. But in practice, theories also have had to "make sense" in terms of ordinary experience. The scientist, looking at nature, draws on more of his human experience than just mathematics. Science is as full of metaphors, good and bad, as is poetry.

Because humans are visually oriented (so much so that the retina is regarded by neurophysiologists as part of the brain), models of nature typically have been visualizable, have made sense to the "mind's eye." This luxury was lost when physics penetrated to the subatomic world, where the "objects" being studied are not much larger than a wave of light and cannot be seen at all. Heisenberg realized that for science to go on, the inherent limitation on the relationship of the scientist to what he or she observed would have to become an acknowledged part of science.

With the fall of visualization came the fall, or at least decline, of causation and force, two concepts that had long been part of science. The logical positivist Rudolph Carnap traced the two doctrines to a common origin. They arose, he said, "as a kind of projection of human experience into the world of nature.

"When a table is pushed, tension is felt in the muscles," Carnap wrote. "When something similar is observed in nature, such as one billiard ball striking another, it is easy to imagine that one ball is having an experience analogous to our experience of pushing the table. The striking ball is the agent. It *does* something to the other ball that makes it move. It is easy to see how men of primitive cultures could suppose that elements in nature were animated, as they themselves were, by souls that willed certain things to happen. This is especially understandable with respect to natural phenomena that cause great harm. A mountain would be blamed for causing a landslide. A tornado would be blamed for damaging a village."

With the advent of twentieth-century physics, both force and causation appear to have been expunged from at least part of science. Einstein did away with the "force" long thought necessary to keep the planets moving in their orbits. For centuries it appeared that something—perhaps God—must push the planets along in their paths. Newton explained planetary motions in terms of a force of universal gravitation but apologized for doing so. In relativity, planets are said to follow their orbits because the orbits describe geodesics, paths of greatest efficiency, in space and time. No force holds them in orbit or pushes them along. In quantum physics causation fails because, given enormous swarms of matter/energy on a tiny scale, the experimenter can predict the behavior of "particles" only in terms of probabilities. Einstein, who helped introduce the statistical approach to physics, felt it was only an expedient and that a cause-and-effect explanation of the behavior of every particle could be worked out if one had sufficient equipment and patience. But Heisenberg showed that a complete cause-and-effect account of subatomic behavior could never be attained by us in the macroscopic world; we are prevented by the uncertainty principle from ever tracing the individual interactions of even a small group of subatomic "particles." Therefore— and this assertion is as much philosophy as science—it is said to be useless to talk about causation operating in a realm where we can never see it or examine it at work. We might as well accept that the probabilities are real.

The decline of determinism may have been made possible in the West by the decline of God's perceived role in the cosmos. For centuries God was assigned the role of prime causer ("unmoved mover," in Aristotle's phrase). Thomas Aquinas was able to construct an impressive proof of the existence of God by placing Him at the head of the cosmic causal chain. With the rise of technology, attention focused on the supposed machinery of nature, and God was increasingly relegated to the role of a creator who set the machinery in motion but now seldom had to look after it, like a successful entrepreneur who need no longer show up regularly at the office. By the beginning of this century, when the machine age began its slow termination, God had withdrawn so far from the scene as to be unable to defend nature against a noncausal philosophy of science.

Although talk of causation survives in modern science, something of the scientific world view has been changed at the root by the quantum principle. Atoms are ubiquitous, and if the subatomic world is nondeterministic, the broader universe must forever present to us an element of chance. It cannot be the precise machine science once portrayed it to be. For many young physicists, the fall of causation was liberating; science had discarded a heavy piece of needless luggage and could proceed afresh. Reading Heisenberg or De Broglie, one gets the sense of adventure that the seventh-century Buddhist Fa-Tsang expressed when he wrote, "Now that we understand that causes are really not causes, any arising will be wonderful."

Einstein, however, did not share in this enthusiasm. He never accepted the contention that because the subatomic world must appear to us in probabilities, it really *is* built upon probabilities. Einstein had been one of the first to demand that the role of the observer be considered in physical theories, but he would not agree that limitation upon observation must be considered the same thing as limitations upon nature. "God does not play dice," he said.

Einstein became isolated from the quantum physics he had helped found. As De Broglie wrote in the early 1950s, "Theoretical physicists are at present divided into two apparently irreconcilable groups. On the one hand, Einstein and his followers are trying to develop general relativity theory, while by far the great majority of theorists, attracted by atomic problems, are trying to develop quantum physics quite independent of general relativity." De Broglie found it "incredible that the two great theories of contemporary physics, the theory of general relativity and that of quanta, should so utterly ignore each other." Max Born wrote of Einstein, "When out of his own work a synthesis of statistical and quantum principles emerged which seemed to be acceptable to almost all physicists, he kept himself aloof and skeptical. Many of us regard this as a tragedy—for him, as he gropes his way in loneliness, and for us who miss our leader and standard-bearer."

As he lived out his final years in a white frame house

on Mercer Street in Princeton, Einstein found himself set apart—enshrined, really—both by his minority position in the philosophy of science and by the international celebrity that had befallen him. His kindly face and rumpled sweaters had become symbols of incomprehensible wisdom. "Because of a peculiar popularity which I have acquired, anything I do is likely to develop into a ridiculous comedy," he wrote an old friend, the Queen Mother of Belgium. To Born he wrote sadly, "In our scientific expectation we have grown antipodes. You believe in God playing dice and I in perfect laws in the world of things existing as real objects, which I try to grasp in a wildly speculative way."

Einstein died on April 18, 1955. A beautiful tribute he wrote for Max Planck, the founder of quantum physics, in 1932, might serve Einstein himself as well:

"Many kinds of men devote themselves to science," Einstein wrote, "and not all for the sake of science herself. There are some who come into her temple because it offers them the opportunity to display their particular talents. To this class of men science is a kind of sport in the practice of which they exult, just as an athlete exults in the exercise of his muscular prowess. There is another class of men who come into the temple to make an offering of their brain pulp in the hope of securing a profitable return. These men are scientists only by the chance of some circumstance which offered itself when making a choice of career. If the attending circumstances had been different, they might have become politicians or captains of business. Should an angel of God descend and drive from the temple of science all those who belong to the categories I have mentioned, I fear the temple would be nearly emptied. But a few worshippers would still remain—some from former times and some from ours. To these latter belongs our Planck. And that is why we love him."

The rift between Einstein and the quantum physicists, like a thousand other less olympian debates, reflected the fact that science is far from achieving a unified view of nature. The very large is assayed via relativity, the very small by quantum physics, the middle ground by a variety of theories of more or less solid reputation. Four "forces"—or "interactions," to use a more modern term—are known to physics: gravity, electromagnetism and two that bind atoms together, the "weak" and "strong" forces. Almost nothing is known about why these interactions exist or why they relate to one another as they do. Why is the electromagnetic force so much more powerful—10^{39} times as powerful—as gravitation? No one knows. There is as yet no theory that encompasses all physical science and explains, in Eddington's phrase, "the wide interrelatedness of things."

In cosmology we see the incompleteness of science magnified. The Big Bang theory is only a sketch, not a painting. It fails to tell us how galaxies, stars and planets formed: If the universe began as a homogeneous soup, why did it not stay so forever? It has little to say about why natural "laws" such as the constant of gravitation should have been dealt out as they were and not some other way; what, for instance, is the role of time in the natural world? It is silent on so fundamental a question as what the cosmos is made of. Atoms? Quanta? The hypothetical building blocks called quarks?

In an effort to close some of the gaps, exotic new cosmologies have been composed to harmonize with the Big Bang theory, while seeking to deepen it.

Several theorists have proposed that the force of gravitation, assumed constant by Newton and Einstein alike, might weaken as time passes and the universe expands. Cosmic history on their terms would resemble a tennis game played on a court that gets slowly larger, while the ball turns lighter and the racquet strings loosen. What justification is there for taking such an idea seriously?

One is Mach's principle. Mach, we recall, proposed that the inertia of any body was the product of its interaction with all the other matter in the cosmos. The young Einstein was struck by the beauty of this idea and also by the strange fact that the *inertial* mass of an object is exactly equal to its *gravitational* mass. In other words, if you weigh something with a spring scale, then set it down on a block of ice or other friction-free surface and pull it sideways with the same scale, you will find that the amount of pull needed to get it moving is the same as its weight. Reasoning that this was unlikely to be coincidental, Einstein assumed that nature was trying to tell him that inertial mass and gravitational mass were in fact identical. This precept became part of the basis of general relativity and was the reason Einstein believed he had incorporated Mach's principle into the theory. At the time he did not know that the universe is expanding—that the distribution of matter, which Mach saw as determining inertia, is thinning out. De Sitter and others subsequently demonstrated that relativity failed to incorporate Mach's principle. It seems at least possible that the failure stems from Einstein's assumption that gravitation remains constant, when instead it may diminish as the universe expands.

In 1937, at age thirty-five, the English physicist Paul Dirac attempted to find, in the relationship between constants of physics, numbers that might reveal something of the basic structure of the world. These were the "dimensionless numbers" that so fascinated Eddington. Dirac chose, as his fundamental unit of time, the interval required for light to cross the radius of a hydrogen atom. The age of the universe, expressed in these units, is slightly under 10^{40}. The coincidence with the ratio between the electromagnetic force and the force of gravitation. 10^{39} suggested to Dirac that the age of the universe had something to do with the force of gravity. Alternately, gravity could be constant and the other numbers be changing, but the Machian associations of gravitation made Dirac favor gravitation as the flexible force.* Gamow, Hoyle and Dicke too proposed at one time or another that the constant of gravitation might be changing. To date, no experiment has confirmed their theories. But the predicted rate of change in the gravitational constant is sufficiently subtle—on the order of five or six parts in 10^{11} per year—to leave ample room for doubt.

---

* Something must remain constant, while something else changes, for the term "change" to have any meaning. To say all physical constants change at the same rate the universe expands is to say that none changes. The situation is reflected in two old philosophical puzzles. One asks, "How do we know but that the universe was formed in toto just last night, complete with all our supposed memories?" The other asks, "What if everything in the cosmos doubled in size overnight? How would we know anything had happened?" On examination, both questions can be seen to be empty by definition. In the first, the expression "last night" is meaningless because time must be transcended to accept the premise as plausible. In the second, space is transcended and so the term "double in size" has no meaning. Things can change in size only if something remains the same to measure the change by.

Another apparently changeless component of nature called into question is the passage of time.

Newton was obliged to make time as grandly inflexible as space. "Absolute, true and mathematical time, of itself, and from its own nature, flows equably without relation to anything external," he wrote. This is time the ultimate monarch, commanding the sky that "impotently rolls as you or I," in the words of Omar Khayyám. But does it make sense to speak of time "without relation to anything external"?

Most of the world's religions and many of its philosophers, among them Spinoza, Kant, Hegel and Schopenhauer, denied that time is real. "To realize the unimportance of time," wrote Bertrand Russell, "is the gate of wisdom." At first sight this position may seem impractical, since we all act as if time were real; Kant was obsessively punctual, and had Spinoza's parents not fled the Spanish Inquisition *in time,* there would have been no Spinoza. The philosophers' point was not that there is no such thing as time but that there is no such *thing* as time; time, they held, is an attribute, like color or harmony, not a fundamental. Events generate time, not the reverse. Lucretius put it this way: "Time does not exist by itself, but from occurrences a feeling arises by which we distinguish what has taken place in the past, what now exists, and what may follow hereafter. We must admit that no one perceives time in itself apart from the motion or immobility of matter." Leibniz wrote, "Instants, considered without the things, are nothing at all. . . ."

Einstein called time a dimension, but if so it is a very strange dimension. We are permitted to move forward or backward in the three dimensions of space, but time flows one way only, toward the future. Eddington called this phenomenon "the arrow of time." What is the reason for it?

Taking a clue from Lucretius and Leibniz—that time cannot be divorced from things—imagine a Big Bang universe "before" expansion began. All matter and energy are compressed to high density. The universe is a chaotic, incoherent swarm of particles. There are no regularly recurring motions by which to measure the passage of time, and no observers to note them if there were. It seems impossible to avoid the conclusion that there can be no time in such a universe. In the classic Big Bang model, this means we cannot inquire about the cosmos prior to the beginning of expansion, cannot know whether the compressed universe existed for an instant or a billion years before expansion began. Indeed, the question is rendered meaningless. It was just that feature of the Big Bang theory that led some cosmologists, Hoyle among them, to reject it.

Now imagine that expansion has begun. In this case we can see a coherent trend in the behavior of the particles: They are thinning out. Two imaginary photographs taken in an expanding universe would show the cosmos at differing densities. In one, the atoms (or galaxies, depending upon at what epoch the ethereal photographer tripped the shutter) are closer together than in the other. Comparing the two, we say, "Ah ha! Time is passing. One of these photographs was taken *after* the other." As Plato said, "Time and the heavens came into being at the same instant."

Notice, though, that while the two photographs will, in themselves, imply that time is passing, they will not tell us in which *direction* it is passing. A motion picture taken before expansion began and another taken after expansion began (the critical reader is asked to pardon the use of "before" and "after") will have in common that either can be run "forward" or "backward" and make equal sense. The preexpansion film will show a chaotic jumble in either case. The postexpansion film will show galaxies either thinning out or coming together, depending upon which way we run it. If we know the universe is "expanding," we can tell the projectionist which is the right way to roll the film, but to say the universe is "expanding" means we have already intuited the arrow of time: Expansion *means* that spatial intervals increase as time passes. This intuition comes naturally to us because we live in the universe, we progress from birth toward death and we are accustomed to the feeling that time flows in one direction only.

So are we back where we started?

Almost. We can see that the direction assumed by the arrow of time may be determined by the dynamics of the cosmos at large—that, in other words, the expansion of the universe reads out in our experience as the fact that time moves forward and not in reverse. But just how does this connection operate? What does the wide universe do to make clocks run down and not wind themselves up, to lead us inexorably from birth to death, to pace us along just as if time were real, if upon considered reflection we are apt to conclude that time is *not* real? Contemplation of the mystery of time requires that we investigate the nature of the substructure that links us with the rest of the universe—which is to say that to understand time, we must seek to understand cosmology.

Several cosmologies envision the arrow of time reversing. Most are closed-universe models in which the present expansion is succeeded by contraction into a fresh fireball, which in turn produces a new universe where time runs backward relative to the one before. Our future is seen as the past of the "next" expanding universe. For time reversal to have meaning, of course, enough information must survive through the fireball so that a trace of the last (and future) cosmos remains. Some have proposed that the laws of thermodynamics may contain information of that sort, but searching for it probably lies beyond the scope of present research. A unique and difficult cosmology constructed by Thomas Gold suggests that the universe oscillates and that time reverses, not in the fireball, but at the point of maximum expansion. When contraction begins, time commences to flow backward. The local effects are at first subtle, later profound—something like those Mach described when he referred to "the famous lobster chained to the bottom of the Lake of Mobrin, whose direful mission, if ever liberated, the poet [August] Kopisch humorously describes as that of a reversal of all the events of the world; the rafters of houses become trees again, cows calves, honey flowers; chickens eggs, and the poet's own poem flows back into his inkstand."

In 1973 Richard Gott proposed an intriguing cosmology incorporating time reversal and antimatter. The existence of antimatter had been hypothesized by Dirac in 1928 on the basis of both relativity and quantum physics, which, he saw, required that each category of known subatomic particles ought to be matched by a corresponding kind of particle with equal mass but opposite charge. If a particle met an antiparticle, Dirac's calculations indicated, the two would annihilate each other in a flash of energy. The prediction seemed so unlikely that Dirac himself called it a "blemish" on his work, but in 1932 a physicist at

Caltech who had never read Dirac's theory discovered, in a cloud chamber, a bit of antimatter. It developed that a complete family of antiparticles does exist, as Dirac had envisioned. Very little antimatter exists on Earth; if it did we would live in a maelstrom. A question for cosmology therefore is whether there are stars and galaxies made of antimatter (present-day observations cannot say, because light from antiatoms looks just like light from atoms) or, if not, why the universe should have been made asymmetrically, favoring matter over antimatter. Gott's contribution to the problem was to picture the Big Bang as generating not one but three universes. The first, Region I, is where we live; time flows "forward" (that is, in the direction we are accustomed to) and matter predominates over antimatter. Region II emerged from the fireball traveling in reverse time; it shares with ours our common genesis, but since their arrows of time point in opposite directions, the two universes have diverged in time ever since, and one cannot observe the other. Region III is a universe of tachyons—theoretical particles that travel faster than light.* The tachyon universe has outrun both other universes since the beginning—they are both limited to the speed of light—and can be observed from neither. Gott's Region II lies permanently in our past, Region III permanently in our future. Both can be confronted only in another fireball. Important to Gott's cosmology is a discovery by the physicist Richard Feynman that antiparticles can be viewed as ordinary particles moving in reverse time. The stray antiparticles we find in our universe could be souvenirs of the stuff that makes up the time-reversed

---

* Relativity theory forbids any particle to be *accelerated* to the speed of light, but permits tachyons because they have *always* been traveling faster than light and will never slow down to it; that tachyons exist has not been established.

universe of Region II, where antimatter physicists study traces of what we consider ordinary matter and, perhaps, wonder why there isn't more of it where they are.

Among the most provocative of the new cosmological theories is John Wheeler's "superspace." Wheeler, a Princeton University physicist who studied under Bohr and taught Dicke, often thought about the rift between relativity and quantum physics, wondering where clues might be found to their reconciliation. A possible answer came to him when he began working on the theory of black holes. Wheeler coined the term "black hole." He thought of them as a meeting ground between general relativity, which predicted their existence, and quantum physics, which governed the world of the very small. What did quantum theory imply about the nature of space on an extremely small scale?

Applying both relativity and quantum physics, Wheeler found that there was a unit of size such that, if a single quantum were smaller than that size, the quantum would act like a tiny black hole. The gravitation of that quantum, though of course minuscule, would be strong enough to wrap space-time around the quantum and remove it from the observable universe. Since nothing smaller than this could possibly be observed, space itself may be considered quantized. The size of the quanta, known as the Planck-Wheeler length, is 10^{-33} centimeter.

Wheeler put it this way: "From an airplane six miles high, the ocean looks smooth. Down at sea level, in a life raft, however, we see that waves are breaking, and the surface is highly irregular; what's more, instead of being merely irregular, there are droplets breaking loose. Now space, too, looks smooth at the scale of everyday life, smooth at the scale of atomic structure and smooth at the scale of nuclear structure. But when one gets down to the scale of distances twenty powers of ten smaller than the scale of nuclear structure, then one predicts that space is foamlike.

"At first this theoretical prediction seems like the most esoteric thing in the world. But recall: There was a time when we thought of a piece of wood as solid; then we learned to think that it's 99.9 percent empty space and that it derives its solidity from electrons whirling around. And now we dare to speak about the possibility that the electrons themselves are, in some sense difficult to be specific about, also made of empty space; or, rather, space that is changing, dynamic, altering from moment to moment."

Owing to the schism between relativity and quantum theory, most modern cosmology has pictured the universe as a relativistic arena all the inhabitants of which—the matter and energy population—were governed not by relativity but by quantum physics. Wheeler's quantizing of space represented an attempt, without abandoning relativity, to bring the arena of space under quantum theory as well. "No principle that we know of in all of physics has the same universal power as the quantum principle," he said. "The more we pursue it, the more it looks as if it is *the* number one principle, and that everything else is, in some way we don't yet understand, derived from it."

The uncertainty principle is broadened in Wheeler's theory to encompass space as well as matter and energy. The cosmological geometry of space is seen as only a *probable* geometry, the sum of the uncertainties of all the quanta in the universe. Determinism withdraws from cosmology; the constitution of natural laws and the behavior of the universe they govern are thought not to have been absolutely determined, but to have arisen quite by chance from a rich menu of options available within the seething fireball of genesis.

Superspace is a hyperdimensional grid Wheeler constructed in which to plot probable pasts and futures of his quantized-space cosmos. It can be represented as a triangular dish. Within it may be traced the evolution of various possible space-time geometries, various cosmoses. The universe of the present is presented as a point, its evolution in time as a line. The walls of the dish represent gravitational collapse. Our cosmic history describes a line departing from one wall and proceeding toward the center of the dish; the line is fuzzed by uncertainty but is hump-shaped in the middle, where the average of all the uncertainties proofs out to approximate the Einsteinian, Newtonian cosmos we see. An oscillating universe wanders in the dish for a while, skids toward one wall, rebounds off it (the Big Bang) and starts a new path of expansion by vote of chance.

The geometry of our universe is only one of many we might have been dealt, in Wheeler's view. There is room in superspace for an infinite number of geometries. "Imagine a gigantic auto body part shop that has parts for every automobile ever made," Wheeler said recently. "Stacked up on the big field are auto fenders of the earliest Ford over in a corner, then as the Ford evolved, the shape of the fender began to change a little, so those altered fenders are stacked a little further along in the field. The man who has the miserable job of locating all the parts has had the wit to place them so that fenders of nearly the same shape lie at nearly the same position, so they can be found easily. This big field would be superspace, and the automobile fenders the various geometries of space."

The cosmic fireball might have dealt a wildly different universe than the one we inhabit, with an alien geometry and different physical laws; many of the universes possible in superspace never form stars, planets or even atoms and molecules. This brings us, by a variant route, to a point we have considered before, that we owe our small existence to that of the broad universe. It suggests, in Wheeler's words, "that there exists a degree of harmony between us and our surroundings that we never realized before. In the past, we looked at our surroundings as if there could be no other, something with which we just had to get along. If this new view is correct, our surroundings are very special and tuned to us, like a plant to its flower: This cycle of the universe like the plant, and we like the flower."

Through all science runs that refrain, that nature is intelligible to us because we in some sense belong to it. The scientist, wrote Einstein, "is astonished to notice how sublime order emerges from what appeared to be chaos." Carl Friedrich von Weizsäcker told his friend Werner Heisenberg, "All our thinking about nature must necessarily move in circles or spirals; for we can only understand nature if we think about her, and we can only think because our brain is built in accordance with nature's laws." James Jeans wondered of the galaxies, "Do their colossal incomprehending masses come nearer to representing the main ultimate reality of the universe, or do we? Are we merely part of the same picture as they, or is it possible that we are part of the artist? Are they perchance only a dream, while we are brain-cells in the mind of the dreamer?" "Ultimately," wrote Mach, "all must form one whole."

To say that nature is comprehensible, that science is

not deluding itself, is an assertion of faith—"reason is one of the articles of faith," said Eddington—but there is nothing wrong in that. After all, we are part of the universe. And the faith of science—that the seamless weave of nature will reveal itself to our reasoned inquiry—that faith is part of the universe too.

GLOSSARY

Astronomical nomenclature can be confusing, especially so when names given in the first flush of discovery later prove inappropriate. Galaxies are still sometimes called "spiral nebulae," though they are not nebulae at all; the original name persists despite the fact that their real nature has been understood for generations. Quasi-stellar objects, or quasars, were so named because they looked something like stars. It has since become evident that they are not stars, and instead may be the cores of forming galaxies, but the name has gained currency and it appears to be too late to change it.

The brief definitions that follow are offered in the hope that they will help in understanding the more technical or idiosyncratic terms that appear in this book.

ABSOLUTE ZERO. The temperature at which a gas has no thermal energy. It equals Minus 459.69 degrees Fahrenheit, or zero degrees Kelvin. (The Kelvin scale is also known as the Absolute scale.) According to measurements of the *cosmic background radiation,* the universe has a temperature of about 2.7 degrees (Kelvin) above absolute zero; if the *Big Bang theory* is correct, this represents the state to which the cosmos, once as hot as billions of degrees or more, has now cooled.

ANTIMATTER. Matter made of particles the same as those of ordinary matter except that their charge is reversed. The proton, for example, has a positive charge, while the antiproton has a negative charge. The term "anti" merely reflects our predilection for the sort of matter we are made of and ought not to be taken as suggesting that nature is necessarily biased against antimatter.

BIG BANG THEORY. The hypothesis that the *expansion of the universe* commenced when the universe—all space, time, matter and energy—was concentrated to high density. Expansion can be said to have begun when the density of the cosmos began to drop, as it continues to do today. The universe is not supposed to have expanded into space; all space was then, as it is now, *in* the universe. If we think of time, the fourth dimension, as having just begun then, we can imagine a three-dimensional universe wrapped up rather tightly in the fourth dimension. The *Big Bang* is an umbrella term for a variety of cosmologies, all having in common that they propose that expansion began violently from a high-density cosmos.

BILLION. The billion employed in this book equals 1,000,000,000, or one thousand million.

BINARY STAR. A system of two stars orbiting a common center of gravity. Such systems are very common; indeed, our solar system came close to forming one. If the planet Jupiter were fifty to a hundred times more massive than it is, thermonuclear processes would take place in its core, it would be a small star and the sun would belong to a binary system.

BLACKBODY. An idealized object that absorbs all radiated energy striking it and so appears black. The concept is useful in physics, where experiments are performed employing an enclosure with blackened interior to approximate a perfect blackbody. The universe as a whole should in theory be a blackbody; for this reason, observations indicating that the *cosmic background radiation* has the characteristics of blackbody radiation are taken as evidence that the background radiation is universal, and probably results from energy released in the *Big Bang*. See *Planck Curve.*

BLACK HOLE. A theoretical object so compressed that its gravity closes space-time around itself, so that no matter or energy can emerge from its surface. A black hole can be regarded as closed off from our universe. Black holes, if they exist, may result from the gravitational collapse of massive stars, as in the aftermath of a *supernova.*

C. Designates radio sources in the Cambridge catalogues of cosmic radio sources. Several editions have appeared—3C for the third catalogue, 4C for the fourth and so on.

CEPHEID VARIABLE STAR. A giant pulsating star whose period of variation is related to its *absolute magnitude.* This quality makes Cepheid variables useful to astronomers as distance indicators. The astronomer observes a Cepheid over a period of months and notes how long it takes to vary in brightness. Its absolute magnitude can then be inferred from its period. Comparing the absolute magnitude with the *apparent magnitude* of the Cepheid in the sky, the astronomer can then say how far away the star is. This method has been invaluable in finding the distances of globular clusters and other galaxies. See *Inverse Square Law.*

COSMIC BACKGROUND RADIATION. Residual energy pervading the entire cosmos, left over from the violent fireball believed to have begun the expansion of the universe. Its existence was proposed as a consequence of the *Big Bang theory*. If the universe began in a Big Bang, the energy radiated by that event should still be present everywhere, though at a level much reduced because it has thinned out with the expansion of the universe. According to the laws of quantum physics, such radiation should conform to a trace of energy against wavelength called the *blackbody* curve. Arno Penzias and Robert Wilson observed what appears to be the cosmic background radiation in 1965; its energy level (often called "temperature") is close to the theoretically predicted value and describes a blackbody curve.

COSMIC RAYS. See *Electromagnetic Radiation.*

COSMOGONY. The study of the origin of the universe, its galaxies, stars and planets. See *Cosmology.*

COSMOLOGICAL CONSTANT. A term introduced into the general relativity equations by Einstein, who later discarded it. It represented an attempt by Einstein to construct a static model of the universe, against the implication of relativity that the cosmos was dynamic instead. The discovery of the *expansion of the universe* eliminated the need for the cosmological constant, though it still shows up occasionally in *cosmology,* if only as a mathematical convenience.

COSMOLOGICAL PRINCIPLE. The assumption that the cosmos at large looks generally the same from all locations— that is, all observers see galaxies in all directions, find them to be rushing apart with the *expansion of the universe* and so forth. The principle serves to affirm that the same physical laws operate throughout the universe—without which assumption it would be very difficult to create cosmological theories—and it bars special pleading, such as arguing that the Milky Way is the center of the universe. In short, our view of the cosmos is said to be typical. See *Perfect Cosmological Principle.*

COSMOLOGY. The science concerned with the nature of the physical universe as a whole. Scientific cosmology involves elements of astronomy, physics, mathematics and philosophy, and cosmological theories normally are created by scientists who work in one or more of those fields of study. Cosmological theories are expected to conform to the laws of physics (or at least most of them), to be consistent with the astronomical evidence and to seem at least marginally reasonable. Philosophical considerations are always present, though more conspicuous in some theories than in others.

DECELERATION PARAMETER. The rate at which the expansion of the universe is slowing down. If, as some observations indicate, the deceleration parameter approximates or equals zero, expansion will go on forever. See *Expansion of the Universe.*

DOPPLER SHIFT. A displacement in the wavelength of light or other radiation caused by the motion of an object, such as a star or galaxy, relative to an observer. If a given galaxy and ours are moving away from each other, the spectral lines in light from that galaxy will be displaced toward the red end of the spectrum. See *Red Shift.*

DWARF STAR. A star of small to medium size and *luminosity*. See *White Dwarf.*

ELECTROMAGNETIC RADIATION. Energy propagated by means of electric and magnetic fields. Its velocity in a vacuum is that of light, and, indeed, light is one form of electromagnetic radiation. Other wavelengths of electromagnetic radiation are described by a variety of names. A tour of the electromagnetic *spectrum,* from long wavelengths to short wavelengths, reads like this: Longwave radio, shortwave radio, VHF television, UHF television, radar, microwaves, infrared light, visible light, ultraviolet light, X rays, gamma rays, cosmic rays.

EXPANSION OF THE UNIVERSE. The motion of galaxies away from one another at velocities proportional to their separation. Another way to put it is to say that the average density of the universe is decreasing, as, according to the *Big Bang theory,* it has been since genesis. The motion of galaxies is inferred from the red shift in their light, for which the only satisfactory explanation known is that they are rushing apart. Expansion pertains not among closely neighboring galaxies—clusters of galaxies like our *Local Group* are bound together gravitationally—but between clusters. See *Doppler Shift, Hubble's Law, Red Shift.*

EXPONENTIAL NOTATION. In order to deal conveniently with the large numbers that pervade astronomy, exponential notation in the power of 10 is used. The rule is that the exponent—the small number to the upper right of the 10—signals how many places to move the decimal point to the right or left. Positive exponents move the decimal point to the right: One million, or 1,000,000, becomes 10^6, while an

American *billion*—the billion used in this book, 1,000,000, 000—is noted as 10^9. Negative exponents move the decimal point to the left: One one-thousandth, or .001, is 10^{-3}. Precise numbers are expressed by multiplication; the mean distance from Earth to the moon in miles, 240,000, can be written 240×10^3, or 24×10^4. The approximate radius of the universe in light-years, 18,000,000,000, becomes 8×10^9.

EXTRAGALACTIC NEBULA. See *Nebula.*

FUSION. See *Nucleus.*

GALACTIC NEBULA. See *Nebula.*

GALAXY. An enormous system of stars, dust and gas. Some are flattened spirals, others are almost spherical, others are irregular splashes of stars. Our sun belongs to the *Milky Way* Galaxy, a large spiral shaped something like a plate or a pinwheel. Galaxies are grouped in clusters, which in turn appear to be organized into vast superclusters. The degree to which this hierarchy pertains in the cosmos—are there clusters of superclusters?—is unknown. The number of galaxies in the universe is estimated at 100 billion, comparable to the number of stars in our galaxy. See *Milky Way.*

GAMMA RAYS. See *Electromagnetic Radiation.*

GLOBULAR CLUSTER. See *Star Cluster, Globular.*

HERTZSPRUNG-RUSSELL DIAGRAM. A plot of the absolute magnitude of stars against their temperature or color. A distinctive pattern emerges, a kind of forking path called the *main sequence.* By diagramming stars that lie both on and off the main sequence, astrophysicists have found valuable clues to stellar evolution and the origins of star clusters. Analysis of the Hertzsprung-Russell diagram helped Walter Baade discover that stars belong to distinct generations, or populations.

HUBBLE'S LAW. The rule that light from distant galaxies is shifted toward the red to a degree directly relating to how far away each is. Also known as the velocity-distance relationship, because velocity is believed to be the means by which the red shifts are generated. See *Doppler Shift, Expansion of the Universe, Red Shift.*

INVERSE SQUARE LAW. The rule that the apparent brightness of an object decreases in proportion to the square of its distance. A star ten light-years away looks one-quarter as bright as the same star five light-years away.

LIGHT. See *Electromagnetic Radiation.*

LIGHT-YEAR. The distance light travels in one year. A light-year equals 5.88×10^{12} miles, or 5,880 billion miles. See *Exponential Notation.*

LOCAL GROUP. A cluster of galaxies that includes the *Milky Way* Galaxy, the large spiral galaxy in Andromeda and perhaps two dozen other systems. The number of galaxies in the Local Group is not known exactly, because the *Milky Way* blocks our view of part of the cosmos.

LOOKBACK TIME. The amount of time light from a remote object, such as a galaxy or quasar, took to reach Earth. The phrase reflects the fact that when we observe across vast distances of space we are also looking back in time; a galaxy a hundred million light-years distant, for instance, appears as it was a hundred million years ago.

LUMINOSITY. The amount of energy a star or other body radiates into space. See *Magnitude.*

M. Designates an object in the *Messier Catalogue.*

MACH'S PRINCIPLE. Broadly stated, an attempt, by Ernst Mach, to express what Arthur Stanley Eddington called "the wide interrelatedness of things." The "weak" form of Mach's principle holds that the motion of one body requires the existence of other bodies against which the motion can be measured; it is meaningless to speak of the "motion" of a single object relative to nothing. The "strong" form of the principle extends this view to argue that not only motion but the inertia of objects depends upon the interrelation of all matter in the cosmos. In both forms Mach's principle influenced a number of physicists, particularly Einstein. However, it is a philosophical assertion, not a scientific one. No mathematical formulation of Mach's principle has appeared, and no one has been able to determine how his purported interrelations of matter propagate themselves.

MAGELLANIC CLOUDS. A pair of irregular galaxies that are satellites of our *Milky Way* Galaxy. They can be seen in the skies of Earth's southern hemisphere, where they resemble detached pieces of the *Milky Way.*

MAGNITUDE. A measure of brightness of astronomical objects. The dimmest star visible to the unaided eye is apparent magnitude 6, while a bright star like Aldebaran is magnitude 1. Still more brilliant objects are designated by 0 or by negative numbers, so that Sirius, the brightest star in Earth's skies, is magnitude –1.6, the full moon –12 and the sun –27. See *Magnitude, Absolute* and *Magnitude, Apparent.*

MAGNITUDE, ABSOLUTE. Intrinsic brightness. Astronomers define it as the *apparent magnitude* a star or other body would have if viewed from a distance of 3.26 light-years. The absolute magnitude of the sun is about 5. Most stars have absolute magnitudes between 0 and 15, though dwarfs as dim as magnitude 20 have been observed, as have a few dazzling giants as bright as –10. See *Magnitude.*

MAGNITUDE, APPARENT. How bright a star or other astronomical object appears in the sky. Apparent magnitude is a function of each star's *absolute magnitude,* how far away it is and whether interstellar dust or gas dims its light in transit. See *Inverse Square Law.*

MAIN SEQUENCE. See *Hertzsprung-Russell Diagram.*

MESSIER CATALOGUE. A list, compiled by the French astronomer Charles Messier, of star clusters, nebulae and other objects in the sky of interest to observers with small telescopes. Some Messier objects belong to our galaxy.

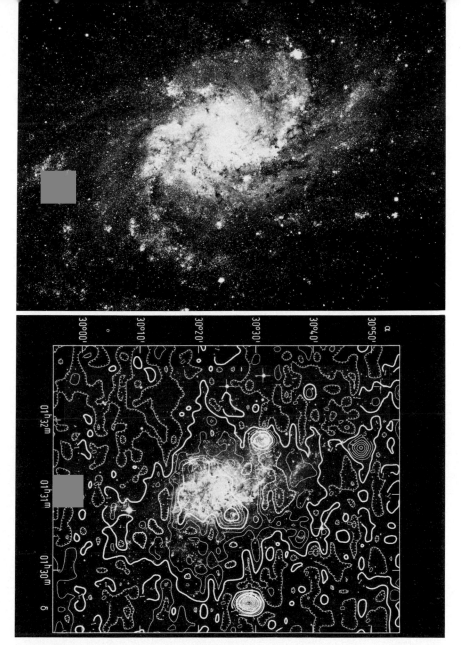

Radio astronomers view the universe in terms of its natural radio energy, just as optical astronomers view it in the wavelengths of light. But in radio, things look much different. *Top:* The galaxy M33, in visual light. *Bottom:* The same galaxy viewed in radio at a wavelength of 6 centimeters, the radio plot superimposed upon an optical photograph. *Photos: M33 optical, Kitt Peak National Observatory; M33 radio, Elly Berkhuijsen, A. von Kapherr, R. Wielebinski*

Grote Reber *(above)*, a radio engineer and amateur astronomer, built the world's first true radio telescope in the backyard of his home in Wheaton, Illinois. Note its striking resemblance to modern radio antennas like the 210-foot dish at the Deep Space Network's Goldstone tracking station in California *(opposite, top)*. *Opposite, bottom:* Computer-generated radio plot of the galaxy Centaurus A (see fourteenth page of third photo section) taken with the Very Large Array of radio telescopes near Socorro, New Mexico. *Photos: Jet Propulsion Laboratory; National Radio Astronomy Observatory*

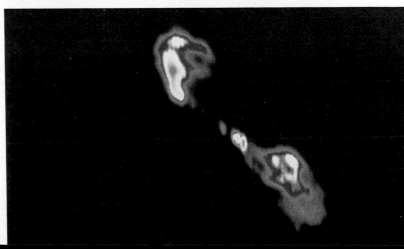

Radio engineer Karl Jansky discovered radio waves coming from space while attempting to trace the origin of noise interfering with long-distance telephone calls. *Above:* Jansky adjusting his rotating antenna, the first to detect cosmic radio energy. *Left:* With a primitive radio map of the northern skies. *Photos: Bell Labs*

George Gamow *(bottom, left)* and his colleagues Ralph Alpher and Robert Herman investigated the physics of a Big Bang genesis in which the expansion of the universe begins in a titanic explosion. They concluded that residual energy from the Big Bang ought to remain throughout the universe today, rather like an echo. Little attention was paid their theory at the time, but nearly twenty years later, in 1965, background radiation with the predicted characteristics was detected by Robert Wilson and Arno Penzias, using a microwave radio horn at Holmdel, N.J. *(top)*. Physicist Robert Dicke of Princeton *(bottom, right)* independently mounted a search for the cosmic background radiation. *Photos: American Institute of Physics; Bell Labs; Princeton University*

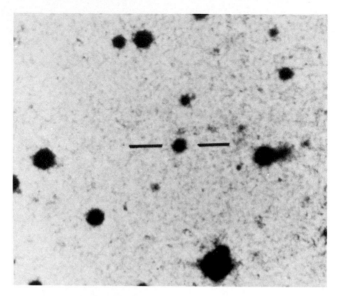

Maarten Schmidt *(bottom, left)* discovered one winter afternoon in 1963 that quasars typically exhibit large red shifts, indicating that they lie billions of light-years away in the expanding universe. He celebrated his discovery with Jesse Greenstein *(bottom, right)*, director of astronomy at Caltech. The highest red shift found by 1982 was that of PKS 2000–330 *(top)*, a quasar estimated to be some twelve billion light-years distant. To be seen at such a remove, quasars must be extraordinarily bright—so bright that some astronomers doubt that quasars really are as remote as their red shifts would lead us to believe. Among the skeptics are physicist Geoffrey Burbidge *(opposite, top, left)* and astronomer Halton Arp *(opposite, top, right). Photos: Caltech (2) Jet Propulsion Laboratory; University of California at San Diego; Halton Arp*

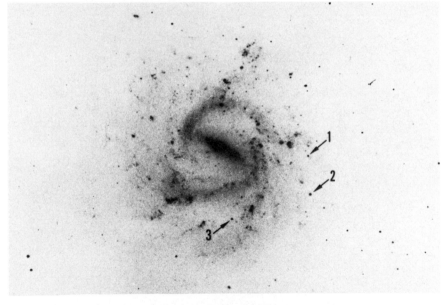

Arp searched the sky near peculiar galaxies, found quasars there, and concluded that the quasars are associated with the galaxies and therefore cannot be farther away than are the relatively low-red-shift galaxies. *Above:* A spiral galaxy and three quasars. Most astronomers believe that the quasars lie deep in the background, billions of light-years behind the galaxies. If Arp's theory of quasars attracted few adherents, his long-exposure photographs revealed many strange and unexplained features of galaxies. *Following pages:* Negative prints of unusual galaxies photographed by Arp. *Photos: Halton Arp, Carnegie Institution of Washington, Mt. Wilson and Las Campanas Observatories*

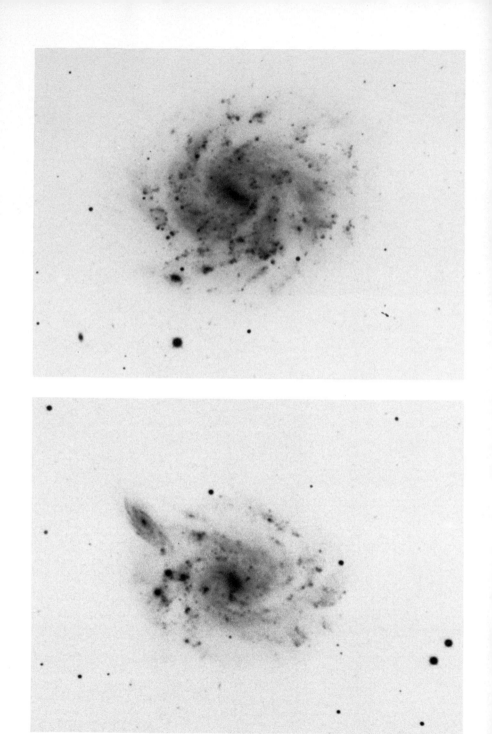

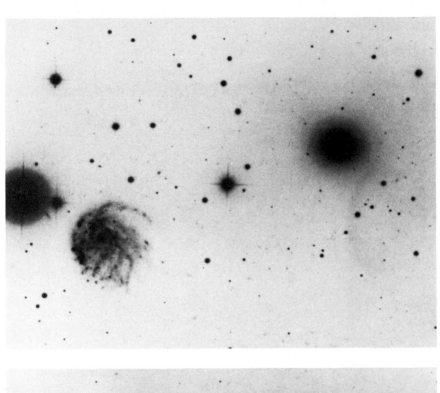

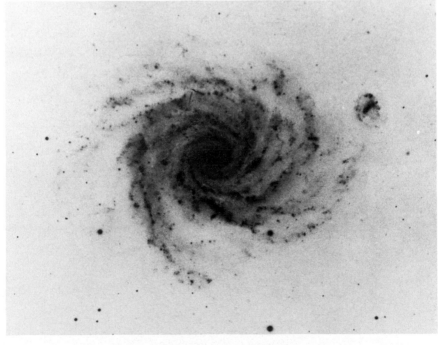

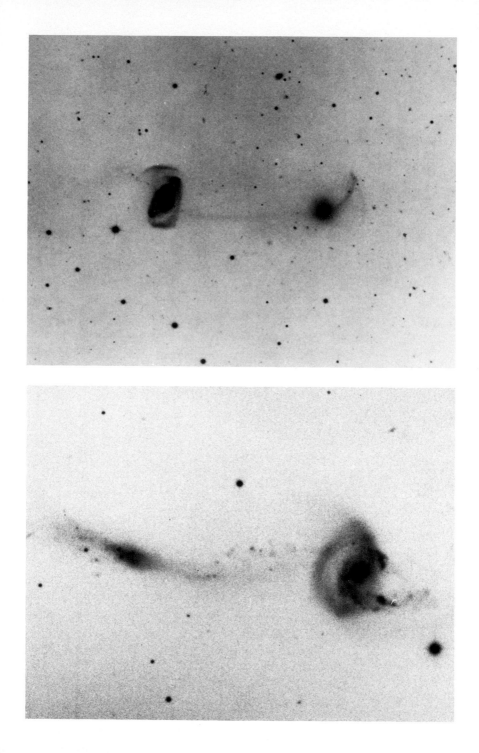

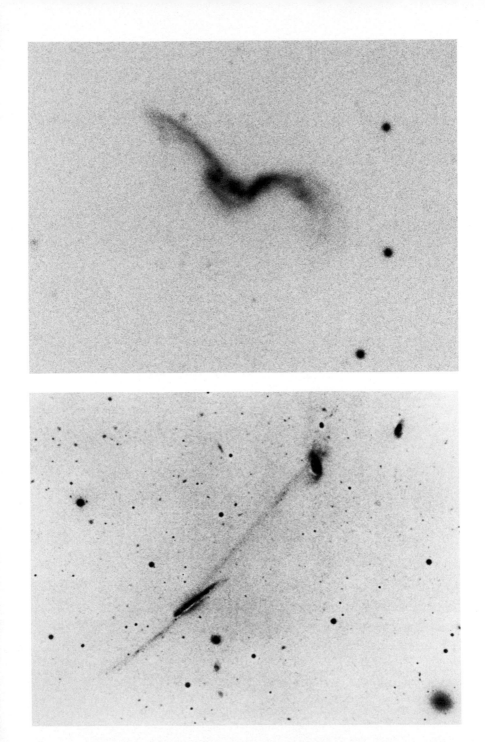

Max Planck *(above)* discovered the quantum principle in 1900. Twenty-seven years later, quantum physicist Werner Heisenberg *(opposite, left)* realized that an element of indeterminacy must be intrinsic to events on the subatomic scale. His "uncertainty principle" liberated the scientific universe from the bondage of strict determinism. Today cosmologists can explore the possibility that the constitution of the universe, and even of natural law, may have resulted from chance events that took place during the first moments of the Big Bang, when the cosmos was the size of a subatomic particle and the uncertainty principle ruled everywhere. *Photos: American Institute of Physics*

John Archibald Wheeler *(right)*, physicist and philosopher, has investigated the role of chance in generating the space-time geometry of the universe. "If this new view is correct," he writes, "our surroundings are very special and tuned to us, like a plant to its flower: This cycle of the universe like the plant, and we like the flower." *Photo: Karl Trappe*

Above: Twin images of a single quasar are produced by an intervening cluster of galaxies (not visible in this electronic exposure) that acts as a "gravitational lens," warping space in its vicinity. *Opposite:* Galaxies outnumber stars in a photograph of a cluster of galaxies taken by Alan Dressler with the 100-inch telescope at Las Campanas Observatory. *Overleaf:* A collision of two galaxies leaves one distorted into a ring 100,000 light-years in diameter. The accident is nonviolent: Few if any stars collide, and the galaxies involved will regain their normal form in a few hundred million years. *Photos: Jerome Kristian, Mt. Wilson and Las Campanas Observatories; Alan Dressler, Carnegie Institution of Washington, Mt. Wilson and Las Campanas Observatories; European Southern Observatory*

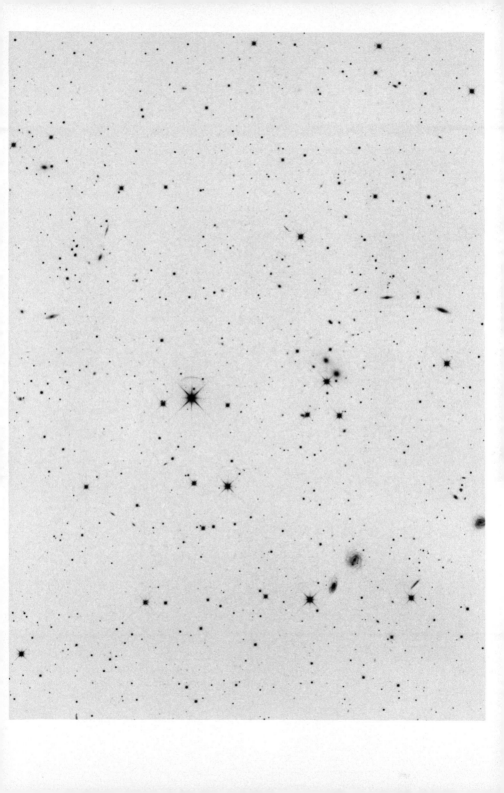

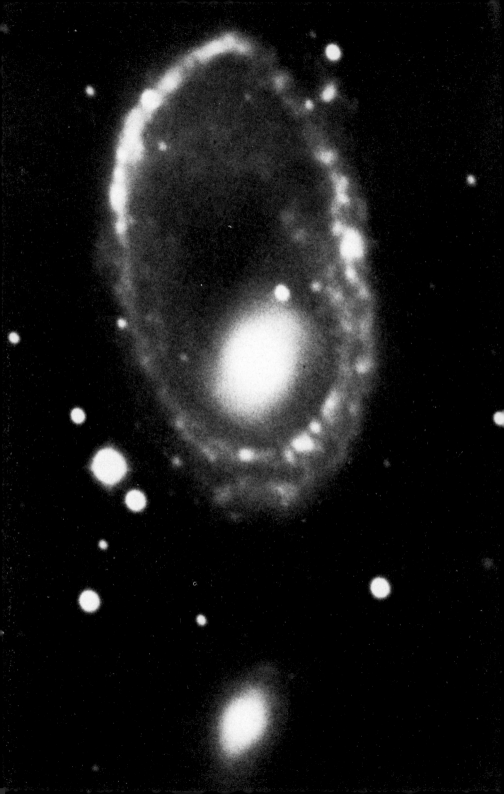

Others, like M31 in Andromeda, are distant galaxies in their own right.

MICROWAVE. See *Electromagnetic Radiation.*

MILKY WAY. A glowing band of soft light in the skies of Earth. It represents the plane of our galaxy seen from within. Because the galaxy is flattened in shape, its stars, dust and gas are much more concentrated along the plane, giving rise to the Milky Way's appearance as a sort of river in the sky. All the individual stars we see with the naked eye, however, both in the Milky Way and elsewhere, belong to our galaxy. See *Milky Way Galaxy.*

MILKY WAY GALAXY. The galaxy to which our sun belongs. It is a flattened spiral perhaps one hundred thousand light-years in diameter; the sun lies about two-thirds of the way out from the center toward one edge. See *Galaxy.*

NGC. Betokens objects listed in the *New General Catalogue,* a list of star clusters, nebulae and galaxies published in 1888.

NEBULA. In the broad sense, almost any astronomical object that looks indistinct in a telescope. *Gaseous nebulae* are clouds of gas within our galaxy; *galactic nebulae* are clouds of gas or dust, some illuminated by nearby stars, others dark and visible only in silhouette; all lie within our galaxy. The so-called *extragalactic nebulae,* or *spiral nebulae,* are other galaxies. The misnomer dates from times before telescopes were large enough to determine the difference between galaxies and clouds in our own galaxy, so that the Andromeda Galaxy, for example, is still sometimes called the Andromeda Nebula.

NEUTRON STAR. An extremely dense object formed from the collapse of a dying star. So named because their gravitational force is great enough to overcome the structure of their atoms, crushing them down to nuclei. See *Nucleus.*

NOVA. Explosion of a star. See *Supernova.*

NUCLEUS. In atomic physics, the central part of an atom, around which swarm electrons. Nuclear *fusion,* the interchange of particles from atomic nuclei in which part of their mass is converted into energy, is the mechanism that makes the stars shine.

PARALLAX. An apparent shift in the position of a star in the sky, caused by a change in the location of its observer. If, for example, a relatively nearby star is photographed from Earth in June and again in December—when Earth has traveled halfway around the sun—its position against more distant background stars will appear to have shifted. What has changed, of course, is our vantage point. (The effect can be illustrated by holding one finger out at arm's length and sighting on it with first one eye, then the other. One eye represents the Earth's position in June, the other in December, and the shift of the finger against background objects is the parallax.) Parallax is used to determine the distances of stars, but because even the nearest stars are quite far away, parallax displacements generally are minor, and the method is accurate out to only a few hundred light-years.

PLANCK CURVE. A tracing of wavelength against intensity of radiation formulated by Max Planck to account for the behavior of *blackbody* radiation. Planck's *quantum* theory correctly predicted the behavior of blackbody radiation where previous theories could not. See *Quantum.*

PLANCK'S CONSTANT. The key expression in Planck's quantum theory of radiation. See *Quantum* and *Blackbody.*

PLANET. A body that shines by reflected light and orbits a star. The sun has nine major planets.

PULSAR. A rapidly spinning *neutron star* that gives off pulses of radio noise.

QUANTUM. A fundamental unit of energy hypothesized by Max Planck to explain the behavior of gas in *blackbody* experiments. The quantum principle portrays energy as released in discrete packets, rather than in a continuous stream.

QUASAR. A compact, bright object which typically is found at great distance, according to measurements of the *red shift* of quasar light. Several theories hold that quasars are galaxies in early stages of formation.

RADIO ASTRONOMY. Observation of the heavens in radio wavelengths. Many astronomical objects, from the sun and the planet Jupiter to giant galaxies, emit radio naturally at various wavelengths and so can be observed with sensitive radio receivers called *radio telescopes.*

RADIO TELESCOPE. A device for receiving radio radiation from galaxies, clouds of interstellar gas and other natural cosmic sources. (No artificially generated radio signals have yet been received from space, except for those transmitted by space vehicles launched from Earth.) Because radio waves are much longer than light waves, radio telescopes generally must be constructed on a larger scale than optical telescopes: Dish antennas of 100 feet or more in diameter are numerous, and the world's largest radio telescope, at Arecibo, Puerto Rico, is 1,000 feet wide.

RADIO WAVES. See *Electromagnetic Radiation.*

RED SHIFT. Displacement of the light received from a star, galaxy or other object toward the lower-frequency end of the *spectrum.* (Displacement toward the *higher*-frequency end of the spectrum is called blue shift.) Red shifts are measured by identifying spectral lines of elements prevalent in the object (such as hydrogen) and comparing their position with a standard spectrum obtained by burning a small sample of the same element in the laboratory or observatory. Red shifts may be produced in several ways, one of which is *Doppler Shift:* The light from, say, a distant galaxy is displaced toward the red because our own and the other galaxy are rushing away from each other, just as a train whistle sounds lower in tone if the train is speeding away. Galaxies do display red shifts, which increase in direct relation to the distance of the galaxy in question; this is primary evidence of the *Expansion of the Universe.* See *Hubble's Law.*

REFLECTING TELESCOPE. A telescope that employs a curved mirror, rather than a lens, to gather light. Most large astronomical telescopes are reflectors.

REFRACTING TELESCOPE. A telescope using a lens to gather light. The classic "spyglass" is a refractor.

RELATIVITY, GENERAL. Einstein's theory of gravitation. See *Relativity, Special,* and *Space-Time.*

RELATIVITY, SPECIAL. Einstein's theory of motion and of the propagation of electromagnetic radiation. The theory can be said to be based upon the idea that the universe harbors no fixed points of reference, no absolute space or time. Measurements of physical behavior must be made from a given observer's frame of reference, and no one frame of reference has priority over another.

SPACE-TIME. A construct invoked in Einstein's theory of *general relativity.* Events in the physical world are viewed as taking place in a four-dimensional continuum, three dimensions of space and one of time. The concept has won widespread acceptance due to relativity theory's virtually flawless record of accurately predicting and accounting for physical phenomena.

SPECTROGRAPH. A spectroscope with a device attached, usually a camera, to record the spectrum for later study. See *Spectrum.*

SPECTROHELIOGRAPH, SPECTROHELIOSCOPE. Instruments for viewing and photographing the sun and its atmosphere in detail. This is made possible by restricting the light observed to a narrow band of the spectrum.

SPECTROSCOPE. An instrument for observing light arrayed in its various wavelengths. See *Spectrum.*

SPECTROSCOPIC BINARY. A *double star* in which the two stars are too close together to be seen as individuals by telescopes on Earth. Spectroscopic analysis of light from the system reveals that it is a binary system, hence the name.

SPECTRUM. Radiated energy broken down into its component wavelengths. In optical astronomy, the energy involved is light. A *spectroscope* spreads the light out in an array according to wavelength, with violet, the shortest wavelength of visible light, at one end, and red, the longest, at the other. A series of lines along the spectrum, generated by a narrow slit in the instrument, signals the presence of various elements under various conditions. These include dark *absorption lines* and bright *emission lines;* by studying them a specialist often can tell a great deal about what a star, planet or galaxy is made of and what conditions prevail there. In studying distant galaxies, measuring the degree to which the spectral lines are displaced toward the red is important, because this is taken as an indication of how far away the galaxies are in the expanding universe. See *Doppler Shift, Hubble's Law, Red Shift.*

SPIRAL NEBULA. A spiral galaxy. See *Galaxy.*

STAR. A self-luminous body of gas sufficiently compressed for thermonuclear processes to operate at the core.

STAR CLUSTER, GLOBULAR. An assembly of stars gathered within a spherical volume of space. The centers of globular clusters are so congested that several stars may be found within a single cubic light-year. Globular cluster stars are generally very old, and many of the clusters are found above and below the planes of spiral galaxies such as our Milky Way. Astrophysicists hypothesize that the clusters formed early in the history of galaxies, when the spirals may have been more nearly spherical in shape and had not yet taken on the flattened form we observe today.

STAR CLUSTER, OPEN. A loose group of stars. Presumably they are generated when a number of stars all form at about the same time out of a common cloud of dust and gas.

STEADY STATE THEORY. The theory that the universe is infinite in space and time, and has always been more or less as it is now. To allow for the *expansion of the universe,* Steady State theorists are obliged to hypothesize that matter is continuously created to fill in gaps left as the galaxies rush apart, so that the density of the universe at large may remain the same.

SUPERNOVA. An exceptionally violent explosion of a star. The detonating star may shine hundreds of millions of times its normal brightness for several days before fading. Supernovae are believed to be the means by which many heavy elements, such as gold and uranium, were created.

TWENTY-ONE-CENTIMETER RADIATION. Radio emissions produced by hydrogen atoms in space. Since most of the matter in interstellar space is hydrogen, study of this natural radio noise has been invaluable in mapping our galaxy and others. See *Radio Astronomy*.

UNCERTAINTY PRINCIPLE. An axiom holding that the location and motion of a subatomic particle cannot both be known with exact accuracy. The experimenter who seeks to know the exact motion of the particle must sacrifice knowledge of its precise position, and vice versa. Therefore a full account of particle behavior can be known only in terms of probabilities.

VARIABLE STAR. A star that changes in *luminosity* regularly. Some variables have periods of less than a day, others weeks, months or even years. Some change dramatically; others so little as to be almost undetectable. See *Cepheid Variable Stars*.

WHITE DWARF. A dense, compact, glowing star believed to be formed from a previously healthy star that exhausted the fuel at its core and collapsed.

X RAYS. See *Electromagnetic Radiation*.

SELECTED
BIBLIOGRAPHY

HISTORY, BIOGRAPHY, MEMOIRS

Among the leading accounts of the development of contemporary astronomy and observational cosmology are *Man Discovers the Galaxies,* by Richard Berendzen, Richard Hart and Daniel Seeley (Science History Publications, 1976), *The Expanding Universe,* by Robert Smith (Cambridge University Press, 1982), *Astronomy of the Twentieth Century,* by Otto Struve and Velta Zebergs (Macmillan, 1962), *The Discovery of Our Galaxy,* by Charles Whitney (Knopf, 1971), and *Quasars, Pulsars, and Black Holes,* by Frederick Golden (Scribner's, 1976). For a broader historical review, see *A History of Astronomy,* by A. Pannekoek (Interscience, 1961), Fred Hoyle's *Astronomy* (Doubleday, 1962), and G. de Vaucouleurs' *Discovery of the Universe* (Macmillan, 1957).

Sound general histories of science include *Science in History,* by J. D. Bernal (Watts, 1954), *A History of Science and Its Relations with Philosophy & Religion,* by William Cecil Dampier (Cambridge University Press, 1949), *A History of the Sciences,* by Stephen F. Mason (Collier, 1962), and *A Short History of Scientific Ideas to 1900,* by Charles Singer (Oxford, 1959).

Biographies and memoirs this book drew upon include George Gamow's *My World Line* (Viking, 1970), Harlow Shapley's *Through Rugged Ways to the Stars* (Scribner's, 1969), and Werner Heisenberg's *Physics and Beyond* (Harper & Row, 1971). Also *Explorer of the Universe: A Biography of George Ellery Hale,* by Helen Wright (Dutton, 1966), *Niels Bohr: The Man, His Science, and the*

World They Changed, by Ruth Moore (Knopf, 1966), *Arthur Stanley Eddington*, by Clive Kilmister (Oxford, 1966) and *Life of Arthur Stanley Eddington*, by A. V. Douglas (Nelson, 1957).

Two books that tell how the 200-inch telescope on Mt. Palomar was built are *The Glass Giant of Palomar*, by David Woodbury (Dodd, Mead, 1939) and *Palomar: The World's Largest Telescope*, by Helen Wright (Macmillan, 1952).

TEXTBOOKS AND INTRODUCTIONS TO ASTRONOMY

There are many enlightening astronomy textbooks, among them George O. Abell's *Exploration of the Universe* (Saunders, 1982), *Essentials of Astronomy*, by Lloyd Motz and Anneta Duveen (Columbia University Press, 1977), Jay M. Pasachoff's *Astronomy: From the Earth to the Universe* (Saunders, 1979), and Pasachoff's *Contemporary Astronomy* (Saunders, 1981). Astrophysics is emphasized in Frank H. Shu's *The Physical Universe* (University Science Books, 1982).

For a comprehensive introduction to astronomy and astrophysics, see *The Cambridge Encyclopaedia of Astronomy*, edited by Simon Mitton (Crown, 1977).

BASIC ASTRONOMY

Introductions written by professional astronomers include *Astronomy*, by Fred Hoyle (Doubleday, 1962), *Astronomy*, by Donald Menzel (Random House, 1970) and *The Universe*, by Otto Struve (MIT, 1962). A small library of books on astronomy have been produced by two popular writers, Patrick Moore in England and Isaac Asimov in the United States; both writers are prolific—Asimov astonishingly so—and rather than attempt to list their books I simply recommend them in general.

Modern developments are discussed in Nigel Calder's *Violent Universe: An Eyewitness Account of the New Astronomy* (Viking, 1969) and John Glasby's *Boundaries of the Universe* (Harvard, 1971), the latter of which is the more technical.

Some of the most beautiful popular writing on astronomy was produced by Arthur Stanley Eddington and his colleague and sometime rival James Jeans. The books are somewhat outdated but still very much worth reading; try Eddington's *The Expanding Universe* (Macmillan, 1933) and *Stars and Atoms* (Yale, 1927), Jeans' *The Universe Around Us* (Cambridge, 1929) and *The Mysterious Universe* (Cambridge, 1930).

Paul Couderc's *The Wider Universe* (Arrow Books, 1960) focuses on extragalactic astronomy, while D. W. Sciama's *The Unity of the Universe* (Doubleday, 1959) touches on philosophical matters, notably Mach's principle; both are straightforward and free from technical jargon.

Two introductions filled with pictures are the *Larousse Encyclopedia of Astronomy* (Prometheus, 1959), and *Pictorial Astronomy*, by Alter, Cleminshaw and Phillips (Crowell, 1969).

For photographs and text concerning galaxies, see *Galaxies*, by Timothy Ferris (Sierra Club Books, 1980; Stewart, Tabori & Chung, 1982) and *The Hubble Atlas of Galaxies*, by Allan Sandage (Carnegie Institution, 1961).

SEMITECHNICAL AND
SPECIALIZED BOOKS

Thirty-three articles on astronomy, ranging from Earth to the quasars, are collected in *Frontiers of Astronomy* (Scientific American/Freeman, 1970). *Beyond the Milky Way* (Macmillan Sky and Telescope Library of Astronomy, Vol. 8, 1969) is slightly more technical but equally engrossing. Two other collections of interest are *Exploring the Universe*, Louise Young, editor (McGraw-Hill, 1963), and *Galaxies and the Universe*, L. Woltjer, editor (Columbia, 1968).

For information about radio astronomy and the astronomies of other parts of the electromagnetic spectrum beyond visible light, see: J. S. Hey, *The Radio Universe* (Oxford, 1971), J. H. Piddington, *Radio Astronomy* (Harper, 1961), H. P. Palmer and R. D. Davies, *Radio Studies of the Universe* (Routledge and Kegan Paul, 1959), John Kraus, *Radio Astronomy* (McGraw-Hill, 1966), A. G. Pacholczyk, *Radio Astrophysics* (Freeman, 1970), I. S. Shklovsky, *Cosmic Radio Waves* (Harvard, 1960), H. P. Palmer and R. D. Davies, editors, *Radio Astronomy Today* (Harvard, 1963), Colin Ronan, *Invisible Astronomy* (Lippincott, 1972) and James Ring, *Infra-red Astronomy* (University of London, 1968).

For information on quasars, see *Quasi-Stellar Objects*, by Geoffrey and Margaret Burbidge (Freeman, 1967), *Quasars*, by F. D. Kahn (Harvard, 1968), *Galaxies, Nuclei, and Quasars*, by Fred Hoyle (Harper & Row, 1965) and an account of the Halton Arp-John Bahcall debate over quasars, *The Redshift Controversy*, George Field, editor (Benjamin, 1973).

Several of the leading figures portrayed in this book left accounts of their work. Edwin Hubble's *The Realm of the Nebulae* (Yale, 1936, 1982) is a classic that demands no expertise on the part of the reader—but remember that Hubble's figures for the distances of galaxies have since been revised upward by a factor of ten. *The Observational Approach to Cosmology* (Oxford, 1937) is another important Hubble volume. Otto Struve's *Stellar Evolution* (Princeton, 1950), more technical, provides an account of the lifetimes of stars written by one of those who contributed most to understanding them. From Walter Baade, who seldom lectured or published, we have *Evolution of Stars and Galaxies* (Harvard, 1963), a transcript of his 1958 Harvard Observatory lectures. For a review of what is currently known about the composition and evolution of stars, see Shklovsky's *Stars: Their Birth, Life, and Death.*

Many of the original papers announcing twentieth-century discoveries of cosmological significance may be found in *A Source Book in Astronomy*, by Harlow Shapley and Helen E. Howarth (McGraw-Hill 1929), *Source Book in Astronomy 1900–1950*, Shapley, editor (Harvard University Press, 1960) and *A Source Book in Astronomy and Astrophysics, 1900–1975*, edited by Kenneth R. Lang and Owen Gingerich (Harvard University Press, 1979).

EINSTEIN AND RELATIVITY

The biography of Einstein that best surveys his scientific research as well as his life is *'Subtle is the Lord . . . ,'* by Abraham Pais (Oxford University Press, 1982). Also recommended is Philipp Frank's *Einstein: His Life and Times* (Knopf, 1970). Ronald Clark's *Einstein: The Life and Times* (World, 1971) is longer and a good source of anecdotal material. Two readable shorter accounts are *Albert Einstein: Creator and Rebel,* by Banesh Hoffman (Viking, 1972), and *Einstein,* by Jeremy Bernstein (Viking, 1973). Ongoing work in relativity and other fields Einstein pioneered is discussed in two collections celebrating the centenary of his birth: *Einstein: A Centenary Volume,* edited by A. P. French (Harvard University Press, 1979) and *Some Strangeness in the Proportion,* edited by Harry Woolf (Addison-Wesley, 1980).

The enduring public fascination with relativity has spawned many popular explanations of the theory. Perhaps the most accessible are *The ABC of Relativity,* by Bertrand Russell (George Allen, 1958), *The Universe and Dr. Einstein,* by Lincoln Barnett (Harper, 1948), *Relativity for the Layman,* by James Coleman (Penguin, 1959) and *Einstein's Universe,* by Nigel Calder (Penguin, 1980), followed closely by Einstein's own *Relativity, The Special and General Theory* (Holt, 1920) and *Relativity and Common Sense,* by Hermann Bondi (Anchor, 1962).

Midway between the nontechnical and the technical levels stand works by two of Einstein's colleagues: *Four Lectures on Relativity and Space,* by Charles Proteus Steinmetz (McGraw-Hill, 1923) and *Einstein's Theory of Relativity,* by Max Born (Dover, 1965).

Gravitation, by Charles Misner, Kip Thorne and John Wheeler (Freeman, 1973), is a massive textbook intended for graduate students in physics, but anyone interested in relativity should at least browse through this remarkable book, if only to have a look at its many biographical sketches, quotations and diagrams. An accurate and easily understood explanation of relativity and gravitational collapse may be found in Walter Sullivan's *Black Holes* (Doubleday, 1979).

Readers interested in contemporary challenges to relativity theory should consult two books by Robert Dicke, *Gravitation and the Universe* (American Philosophical Society, 1970) and *The Theoretical Significance of Experimental Relativity* (Gordon & Breach, 1964). The former is the less demanding.

Relativity is placed in the context of the physics of its day by A. D'Abro in his *The Rise of the New Physics* (Van Nostrand, 1939; Dover, 1951) and analyzed in terms of psychological context by Lewis Feuer in *Einstein and the Generations of Science* (Basic Books, 1974).

COSMOLOGY

Cosmology is a highly theoretical subject and the non-mathematical reader is best advised to read several works to gain a proper perspective. On the general level we have Edward R. Harrison's erudite *Cosmology: The Science of the Universe* (Cambridge University Press, 1981), Jagjit Singh's rewarding *Great Ideas and Theories of Modern Cosmology* (Dover, 1961) and a charming book by F. P. Dickson, *The Bowl of Night: The Physical Universe and Scientific Thought* (M.I.T., 1968). The perspective of an advocate of the Big Bang theory is supplied in non-scientific terms by George Gamow's *The Creation of the Universe* (Viking, 1952), that of the Steady State in Fred Hoyle's *The Nature of the Universe* (Oxford, 1960). *Theories of the Universe,* edited by Milton Munitz (Macmillan, 1957) offers excerpts from the writings of cosmologists from Lucretius through De Sitter, Lemaître, Milne, Gamow, Bondi and Hoyle. *The Universe and Its Origin,* by H. Messel and S. T. Butler (W. & J. Mackay, 1964) offers papers by several contemporary cosmologists, as do

The Emerging Universe, William Saslaw, editor (Virginia, 1972) and *Cosmology, Fusion & Other Matters,* Frederick Reines, editor (Colorado, 1972), a memorial volume issued in tribute to Gamow.

Hermann Bondi's *Cosmology* (Cambridge, 1960) is a standard reference work, and semitechnical. Other helpful books on this level include *Fact and Theory in Cosmology,* by G. C. McVittie, *The Origin and Evolution of the Universe,* by Evry Schatzman (Basic Books, 1965) and *The Structure and Evolution of the Universe,* by G. J. Whitrow (Hutchinson, 1959).

J. D. North's *The Measure of the Universe* (Oxford, 1965), an extensive survey of modern cosmology, is recommended to advanced students. For cosmological writings by leading theorists themselves, see Bondi's *Rival Theories of Cosmology* (Oxford, 1960), James Jeans' *Astronomy and Cosmogony* (Dover, 1961), De Sitter's *Kosmos* (Harvard, 1932), Hubble's *The Observational Approach to Cosmology* (Oxford, 1937), Hannes Alfvén's *Worlds-Antiworlds* (Freeman, 1966), Georges Lemaître's *The Primeval Atom* (Van Nostrand, 1950) and P. J. E. Peebles' *Physical Cosmology* (Princeton, 1971). Those interested in Immanuel Kant's contributions to cosmology should note that a translation of his *Universal Natural History and Theory of the Heavens* has been published by the University of Michigan Press in unabridged form.

In recent years theoretical-particle physicists have been investigating how the nature of matter and energy in the universe today, and even the constitution of the natural laws that dictate their behavior, may have been determined by quantum events that occurred in the first microseconds of the Big Bang. These developments are reviewed in two superior books, *The First Three Minutes,* by Steven Weinberg (Bantam, 1979) and *The Cosmic Code,* by Heinz R. Pagels (Simon & Schuster, 1982).

PHILOSOPHY OF NATURE

These are some of the sources drawn upon for the philosophical issues raised in the final chapter:
Bohr, Niels, *Atomic Theory and the Description of Nature* (Cambridge, 1934).
Bondi, Hermann, *Assumption and Myth in Modern Theories of Science* (Cambridge, 1967).
Burtt, E. A., *Metaphysical Foundations of Modern Physical Science* (Harcourt, 1927).
Carnap, Rudolf, *Philosophical Foundations of Physics: An Introduction to the Philosophy of Science* (Basic Books, 1966).
D'Abro, A., *The Evolution of Scientific Thought From Newton to Einstein* (Dover, 1950).
Eddington, Arthur Stanley, *Fundamental Theory* (Cambridge, 1946).
——,*The Philosophy of Physical Science* (Cambridge, 1939).
Gold, Thomas (ed.), with Schumacher, D. L., *The Nature of Time* (Cornell, 1963).
Heisenberg, Werner, *Across the Frontiers* (Harper & Row, 1974).
——, *Physics and Philosophy* (Harper, 1958).
Hubble, Edwin, *The Nature of Science and Other Lectures* (Huntington Library, 1954).
Jeans, James, *Physics and Philosophy* (Ann Arbor, 1958).
Planck, Max, *Scientific Autobiography and Other Papers* (Philosophical Library, 1949).
——, *The New Science* (Meridian Books, 1959).

Poincaré, Henri, *The Foundations of Science* (Science Press, 1913).

Reichenbach, Hans, *Modern Philosophy of Science* (Humanities Press, 1959).

———, *The Philosophy of Space and Time* (Dover, 1957).

Schilpp, Paul (ed.), *Albert Einstein: Philosopher Scientist* (Library of Living Philosophers, 1949).

Schlegel, Richard, *Time and the Physical World* (Dover, 1958).

Toulmin, Stephen (ed.), *Physical Reality* (Harper & Row, 1970).

Weyl, Hermann, *Philosophy of Mathematics and Natural Science* (Princeton, 1949).

Whitehead, Alfred North, *Essays in Science and Philosophy* (Philosophical Library, 1947).

———, *Modes of Thought* (Macmillan, 1938).

———, *Science and the Modern World* (Macmillan, 1925).

Whitrow, G. J., *Natural Philosophy of Time* (Nelson, 1961).

INDEX

absolute magnitude:
  of Cepheids, 34–35, 36
  Herschel's use of, 31
  Hertzsprung-Russell diagram and, 100
Adams, W. S., 151
alcohol, in space, 151
Alexander the Great, 63–64
Alpher, Ralph, 130, 143, 148, 153, 154
  Alpher-Bethe-Gamow theory and,
    127–128, 129
  Gamow and, 127–128, 130
Alpher-Bethe-Gamow theory, 127–129
American Association for the Advancement of Science, 191–192
American Astronomical Society, conference of (1924), 50
Anderson, John, 107
Andromeda Galaxy (M31):
  Baade's study of, 103–106, 107–108
  Cepheids in, 50
  Hubble on, 49, 51–52
  location in sky of, 26
  Lundmark on distance of, 49
  Öpik on distance of, 49
  relative size of, 93
  Sandage on distance of, 110

Shapley on distance of, 108
supernovae in (1885), 30–31, 41–42
Annals (Harvard Observatory), 32
antimatter, 238–239
Aquinas, Thomas, 229
Arcturus (star), 26, 54
Arequipa (Peru), observatory, 32
Aristotle, 64, 229
Arp, Halton:
  Bahcall vs., 191–193
  Big Bang theory attacked by, 190
  at Palomar, 188, 198–199
  on quasars, 188–193
  Sandage and, 188–189, 192, 193–199
asteroids, origin of, 85
Astrophysical Journal, 47, 94, 105, 138
Astrophysical Journal Letters, 149, 177–178
Atkinson, Robert, 101
Atlas of Peculiar Galaxies (Arp), 188, 189
atoms:
  Mach on, 65
  neutron capture cross-section of, 127–128, 129
  recycling of, 126–127

277

278   *Index*

atoms: (*cont.*)
    strong force in, 121, 232
    weak force in, 121, 232

Baade, Walter, 122, 138, 163
    Andromeda Galaxy studied by,
        103–106, 107–108
    Cepheids studied by, 108–110, 203
    on exploding stars, 211
    Gamow and, 125–126
    at Mt. Wilson, 102–106
    with Palomar 200-inch telescope,
        108–109
    on varying ages of stars, 109–110
Bahcall, John, 199–200
    Arp vs., 191–193
Bahcall, Neta, 191
Bailey, Solomon, 37
Baum, William, 108
Berkeley, George, 64, 65–66
Bernard, Edward E., 51
Beta Andromeda (star), 26
Bethe, Hans, 101, 128, 139
Big Bang theory:
    Alpher-Bethe-Gamow theory and,
        127–129
    Arp's attack on, 190
    background radiation and, 141–154
    coinage of, 125
    Gamow's work on, 125, 127–131
    history of science and, 220–221, 232–
        233
    rationality of, 162–163
    Steady State theory vs., 162–163
    theory vs. observation in, 164–165
    time and, 236–237
binary stars, eclipsing, 33, 35–36
blackbody radiation curves, 148, 150–
    151, 152, 222
black holes, 102, 209–216, 240
Blake, William, 133
BL Lacertae objects, 184
blue shift, 54
Bohr, Niels, 139, 224, 226–227
Bolton, John, 179
Bolyai, János, 69
Bondi, Herman, 157–161
Boötes (constellation), 58, 59

Born, Max, 76, 230
Bunsen, Robert, 29
Burbidge, E. Margaret, 162, 181, 182, 199
Burbidge, Geoffrey, 162, 187, 199
Burke, Bernard, 148–149

Cannon, Annie Jump, 32–33
carbon monoxide, in space, 151
Carnap, Rudolph, 227–228
Carswell, R. F., 182
Cassiopeia (constellation), sighting of, 26
causation:
    Einstein on, 228
    probability and, 227–229
Cayley, Arthur, 69
centripetal force, 66–67
Cepheid variable stars:
    in Andromeda, 49–50, 51
    Baade's study of, 108–110, 203
    as indicators of distance, 36–37, 106, 108, 204
    measuring distances to, 33–35
    in NGC 6822, 51
    in spiral nebulae, 49–50
Chandrasekhar, Subrahamanyan, 210–211
Cheseaux, J.P.L. de, 87
Chuang-Tzu, 23
Clark, Alvan, 96
Coma Berenices (constellation), 58, 107
comets, origin of, 85
complementarity, principle of, 226–227
"Concerning the Quantum Theory of the Absolute Zero of Temperature," 123
Contributions (Mt. Wilson), 39
Copernicus, Nicolaus, 40, 220–221
Corning Glass Company, 98–99
cosmic abundance curve, 127, 128
cosmic background radiation, 142–153
cosmological constant, 78, 79, 119
cosmological principle, 160–161
Cottingham, E. T., 75
Creation of the Universe, The (Gamow), 153
Critchfield, C. L., 101